MW01331544

WISDOM AT A COST

A STORY OF SURVIVAL, STRENGTH, AND LEADERSHIP

BY JOHN MATHEWS

Dedication

To my dearest Ashlynn,

As you navigate your own journey into womanhood, always remember to *"Stay the Course."* You are my greatest inspiration and proudest achievement. This book is for you. May it remind you that resilience and courage can overcome any challenge.

PREFACE

Like many, I see myself as an honest person. However, if you think about it, being honest isn't difficult. You only need to be honest when someone asks you a question. That's the key to honesty: you don't have to share any information beyond what is asked. Being transparent, on the other hand, is something completely different.

Transparency means sharing information openly without being asked. As honest as I thought I was or may have appeared, I was never truly transparent. Let's face it: there's nothing mysterious about transparency—it's not alluring. Who walks around revealing everything to everyone? Until now, I've kept my story private, sharing it only with a select few. This book represents my effort to be transparent, raw, uncut, and vulnerable.

Like everyone else, I've experienced the full range of human emotions—happiness, sadness, fear, pleasure, anger, guilt, shame, regret, and more—with shame being the most challenging. It's the kind of feeling that burrows deep, settling into the core of who you are. We all carry it; some try to bury it with work, alcohol, drugs, relationships, or a combination of them—only to find themselves feeling even emptier than before.

Shame is what prevented me from being transparent. It took years before I had the strength to share my story with someone other than my family. Even then, I couldn't tell the story without getting emotional. After all, I buried those thoughts deep to survive. I didn't know what would happen if I let them out. The last time I allowed those thoughts to run freely through my mind, it created a man I couldn't control. A man who was broken. A man with so much shame I couldn't even look him in the mirror. It was a reflection I couldn't stand.

I have found that we all have shame, and it comes in all different shapes and sizes. Over the years, some of my closest friends have shared some of their most shameful stories, and I feel blessed to have heard each of them. In that moment, you bond on a deeper level and understand that we are not all that different. In this book, I speak openly. I will share the stories that shaped me into who I am today.

You are going to embark on a journey, my journey. I walk you through what it feels like to be a broken boy with track marks on his arms to one of the leaders of the largest disaster the state of Tennessee has ever had. The journey is full of shame, trauma, and disappointment, but more importantly, it is filled with resiliency, happiness, and success. I will share the wisdom I have acquired over the years. The wisdom I gained from the decisions that led me from incarceration to becoming the Vice President of a 150-million-dollar company.

All life lessons come at a cost. Wisdom comes at a cost. Yet it is something you cannot buy; it's earned. It's earned at your expense and that of others. It comes at a cost that you will have to pay whether you want to learn the lesson or not. There are no discounts or Black Friday sales. You will always pay in full, with a hidden interest rate that will come due when you least expect it. Some lessons come at a cost that is too much to endure. These are the ones that are the hardest to learn. Like the death of a child or a spouse. Lessons that only certain people will understand.

The wisdom I gained came from years of drug abuse, which caused both mental and physical damage to my mind and body. I suffered years of abuse from an alcoholic father who never healed from his childhood trauma. And then, I managed the 2016 Gatlinburg Wildfires, the largest disaster in the state of Tennessee, where 14 people lost their lives. All of this came at a cost.

This book is not a step-by-step guide to becoming sober. It isn't filled with self-help tools to "fix" your problems. It was hard enough to fix my own problems, let alone others. Instead, this book is filled with stories of resilience and how I healed from a life full of trauma. It's a story of how I hit rock bottom and climbed my way out. It is a story about the cost of wisdom.

And in sharing it, I hope you find a piece of yourself within these pages. Whether you have faced addiction, adversity, or loss, I want you to know this: You are not alone.

The past does not define you. You are stronger than you think. This is my story. But perhaps, in some way, it is yours too.

NOTE FROM THE AUTHOR

To respect the privacy of certain individuals, some names and details have been altered.

CHAPTER 1

HOW THE FUCK DID I GET HERE

I shut the bathroom stall door, and the palms of my hands start to sweat as my breathing quickens. My chest tightens, and all I can hear is my heart pounding in my ears, louder than anything else around me. The uncontrollable sound of my thoughts soon takes over, spiraling faster than I can stop them.

I pace back and forth in the small space, staring down at my feet as beads of sweat drip from my forehead. I loosen my tie, desperate to feel like I can breathe again. The walls of the stall close in around me. Maybe this is what a panic attack feels like.

I put my head in my hands, *How the fuck did I get here?* The thought loops endlessly in my mind as the weight of everything I've lived through presses down on me. Then, something clicks—*You've been through so much to get here,* I tell myself. *Get your shit together!*

The memories of my life flash through my mind like a vivid and inescapable slideshow. Images of sleeping in my car the first night my dad kicked me out. The cold, hard concrete of a county jail cell pressed against my back. Puncture bruises lining the insides of my arms as the lifeless eyes of a junkie stare back at me in the mirror. Each memory hits like a gut punch, freezing me between who I was and who I'm supposed to be.

I push open the stall door and stumble toward the sink. The cold splash of water on my face jolts me back to the present. I lift my head and meet the gaze of the man staring back at me in the mirror. It's a face I haven't seen in a long time—uncertain, uncomfortable, and completely exposed. It's the look of someone still trying to piece it all together.

The distance from the bathroom door to the podium is no more than 50 feet, but it feels like an eternity. Each step is slow and heavy, like I am wading through water, dragging the weight of every mistake I've ever made behind me. I'm moving, but I feel like I'm outside my body, watching myself from somewhere far away.

My mind races as I think about the sea of reporters waiting just beyond the door, their cameras and notebooks aimed and ready as I prepare to comment on the deaths caused by the wildfire. *What can I possibly say? How will I find the words?*

The remarks I'm about to speak aren't just words—they're stories carved into my bones, the lessons I've lived through and barely survived. Standing there, staring into the rows of faces before me, I know this is no ordinary moment.

This isn't just about answering questions; it's about the journey that brought me to this exact spot.

As I grasp the side of the podium, I adjust the microphone and look out at the reporters. My voice steadies as I begin. The memories aren't just flashbacks anymore—they're proof of the cost of survival and the resilience that carried me here. And now, standing here in the aftermath, news reporters focus on me as I respond to questions about the 14 deaths of our neighbors and visitors.

CHAPTER 2

THE SCAR THAT STAYED

In the 1980s, my hometown of Broward County, Florida, was full of cheap land and colorful individuals. Twenty-eight miles north of Miami, my family's home sat a few blocks east of Everglades National Park, the endless swamp that covers nearly 4,200 square miles of southern Florida. The only road that gets someone from east to west across the Everglades is known as Alligator Alley because that's about the only living thing you'll find once you pass Broward County.

The land we lived on was so inexpensive that we were among the few families in the neighborhood brave enough to own a house there. When it rained, the backyard would flood with two feet of water—deep enough to pull us on a kneeboard tied to the back of our neighbor's three-wheeler, as if we were skating across a lake. I can still remember the smell of the damp, humid air and the sound of cicadas playing like a never-ending symphony of the Everglades. This was the world we grew up in— where the swamp wasn't just home, but an endless playground shaped by water, wildlife, and the wild imaginations of those who dared to call it theirs.

Mom and Dad couldn't afford to take our family out to dinner often, but on special occasions, we would go to this little Italian restaurant a few miles away from our house. Mom said it was the closest thing on this side of the Atlantic Ocean to getting an authentic Italian meal like the ones from her childhood.

The restaurant served food in a traditional family style, using large white ceramic dishes baked in the oven. My earliest childhood memory is of this place. I was just three years old, celebrating my six-year-old sister Leslie's birthday. My twin brother, Mike, sat to my left, while Mom and Leslie sat across from me. Dad couldn't join us because he was at work.

After the server placed the food on the table, Mom began to cut into the lasagna. In a sudden and terrifying moment, the dish slipped from her hands and fell onto my sister's lap. The plate was scorching hot, and Mom immediately rushed to help, fearing it would burn Leslie's leg. Meanwhile, amidst the chaos, some of the hot food had landed on my face, and I found myself screaming in agony.

I vividly remember the sensation as if it were yesterday: I turned around and grabbed the red pleather cushion on the back of the booth, pressing my face into the fabric in an attempt to soothe the searing pain. My skin was peeling off from the heat of the food. "It burns; it burns!" I screamed over and over again.

I could not open my eyes, but I could feel my body being lifted out of my chair, carried several hundred feet, and then placed into the front seat of my mom's blue Buick. Since I was never allowed in the front seat, I knew something was wrong. Mom was driving with one hand and holding a bag of ice to my face with the other. The sunlight outside was so bright that it burned my eyes when I attempted to open them and look out the window.

We went straight to our pediatrician's office. I was in so much pain that it took both the doctor and Mom to hold me down while they removed the burnt skin from my face. He then bandaged my entire head but left an opening just large enough so I could see. I could hear him expressing his concerns about the potential long-term effects of the burn to Mom. Although I can't recall the specifics of their conversation, I imagine it wasn't positive.

At home, Mom carried me to my bed. When she thought I was asleep, I heard her praying over me, "Lord, please don't let my baby be deformed." As she prayed that night, I could see the determination in her eyes—the quiet strength of a mother who wouldn't let her child face the world alone.

Her prayer was answered. The burn eventually healed, but the scar remained—a silent reminder that trauma doesn't just disappear; it lingers, shaping who we are and how we grow. That scar revealed something I couldn't have understood then: our capacity to heal, even when the pain feels overwhelming. It taught me that adversity, whether small or large, leaves an imprint that stays with us, often in ways we don't immediately understand.

That day in the red booth marked my first encounter with the weight of resilience. The pain had faded, but the lesson endured: *scars tell our stories, and this one marked the beginning of mine*. That scar didn't just mark my skin; it reminded me that survival isn't about avoiding pain—it's about facing it head-on and letting it shape who you become.

Over time, I began to realize that not all scars are physical. The emotional scars we carry often come from a deeper cut, and many of mine weren't even my own—they were passed down to me, a legacy of trauma from those who came before me.

CHAPTER 3

GENERATIONAL TRAUMA

Being part of a family carries a weight—one made heavier by the trauma and patterns passed down through generations. In my family, the scars were not just physical but emotional, etched into my very identity like a birthright.

What began as my grandparents' struggles with abuse and survival cascaded through the years, shaping not only their lives but also those of my parents, their siblings, and, ultimately, me. Inheriting trauma has a way of lingering, weaving itself into the fabric of who we are, even when we try to escape it.

My mother is a first-generation Italian American. She was born in New Jersey and moved to South Florida with her parents and brother when she was a little girl. Her father, my grandfather, arrived at Ellis Island in 1946 at age twenty-one with his seventy-year-old mother in tow. Neither of them spoke a lick of English. They had left their small town in Italy, the only home they had ever known, all for the promise of the "American Dream."

When my grandfather and his mother stepped off the ship and reached the front of the customs line, he told the officer his name: "Pasquale Serpico." As he watched the man inscribe it into the immigration ledger, a slow smile crept across his face. In that moment, he knew—a new life was beginning, and there was no looking back.

Pasquale settled in Brooklyn, NY, and quickly found a low-paying labor job. Immigrants like my grandfather were in high demand in the labor force because their employers knew they could pay them less than other trade workers and enlist them to do the kind of back-breaking work that other workers simply didn't want to do.

Aside from low pay and long hours, they were often treated like outcasts. Being spit on and harassed was his regular penance for not being American. Yet he endured it all, driven by dreams of creating a better future for his family—a future he knew he might not live to see himself.

When he met my grandmother, she was only 16. She was a hard-nosed, forceful woman who was also Italian but born in America. Their marriage wasn't built on romance—it was built on survival. They were both trying to escape their own forms of hardship, finding a way to forge a future together out of necessity. Three years later, in 1949, after she turned eighteen, he had punched his ticket both in love and in citizenship.

Together, they moved upstairs into his in-laws' East Paterson, New Jersey, home. They had two children: my mother, Patricia, and her brother, Joseph. Pasquale's parenting style was exceedingly strict and borderline abusive. He ran the house with an iron fist, a controlling and often menacing presence. Pasquale believed discipline was survival, but his inability to show affection left emotional scars on my mother that she carried long after she moved out.

The line of bad men in my mother's life didn't stop with her father. Her uncle and grandfather, who both lived downstairs in the East Paterson home, sexually abused her. When no one was looking, they found any opportunity to get my mother alone, and when they did, each would molest her- separately. This went on for years, right under the nose of the rest of the family.

If anyone knew what happened or what monsters her uncle and grandfather really were, nobody ever said. My mother never told her parents what they did to her, but shame and fear would follow her for the rest of her life.

Mom found her way out of her family and home life at 19 when she met my father, Bill Mathews. He was nine years older, had already been married once, and, technically, was still married when they met. Whether their relationship sped up the inevitable or simply gave him an excuse to move on, it didn't take long for his first marriage to end.

For Mom, he wasn't just a partner—he was an escape. I can't say that love brought them together, but convenience certainly did. The idea of escape wasn't just Mom's story—it ran deep in Dad's bloodline too. While she was running from a troubled home, he was shaped by a legacy of men who ran from theirs. He never met his father, Wilford Sparks, because his old man skipped town one day without a word, leaving his wife Clara with their two young children.

The pattern of abandonment in Dad's family didn't start or end with Wilford Sparks. It stretched further back, woven into the fabric of his lineage—generation after generation of men who chose to leave rather than stay.

My great-grandfather, James L. Seale, was no different. He walked away from his family chasing a dream, setting his sights on becoming a Texas Ranger. Deserter, though he may have been, he at least achieved his dream. Years ago, our family managed to track down his last known whereabouts and found his name in a book titled America's Greatest Gunslingers, along with the Colt .45-caliber pistol issued to him by the Texas Rangers. We discovered that he used that same gun in a duel against another Texas Ranger who challenged him for reasons unknown. Duels were still legal in Texas at that time. During the duel, he suffered a shot to the hip but killed his opponent. After sustaining the injury, he received an honorable discharge from the Rangers and lived the rest of his life in Texas.

While the men in Dad's family had a history of leaving, the women were left behind to pick up the pieces. His mother was no exception. As a single parent, she worked long hours as a waitress to provide for her two kids. Dad said there were many times when she would drop him and his sister off at the nearby park with a fishing pole. They would fish all day until she finished her shift and picked them up. While Dad and his sister spent their days fishing at the park, their mother spent hers trying to build a better future for them. But survival often came with compromises, and before long, Clara found companionship with a regular customer at the diner—Preston "Ted" Mathews. Soon, my dad and his sister were calling him stepdad.

It turns out Ted (or Colonel Mathews, as Dad was required to call him, though the man never served a day in the military) wasn't much better than my dad's old man. Generational trauma has a way of digging itself into the foundation of a family, cycling through the years until someone decides to stop it.

My dad's sister, Sonja, my aunt, bore the same scars my mother had endured when her new stepdad, Ted, started sexually abusing her. Like my own mother, my aunt kept Ted's secret. Unfortunately, there's an understanding that develops in families when abuse occurs, and no one has the courage to confront it.

Ted relocated the family to St. Simon's Island, Georgia, but when Dad turned eighteen, he escaped the dysfunctional home by enlisting in the Coast Guard. He served four years and never returned home.

During his military career, he met his first wife, Peggy. They married quickly and had a daughter, Heather, who is ten years older than me, almost to the day.

Five years after marrying Peggy, Dad repeated the sins of his own father and grandfather, leaving his wife and young child. Dad hardly ever mentioned that he had another child outside me, my brother, and my sister. My entire life, I would have to correct him during conversations with friends or co-workers, when he would tell them about his "three kids." I would have to remind him about Heather saying, "No, Dad, you have four kids, remember?" But he would shrug his shoulders and shake it off as though it was an easy mistake, just an oversight.

I believed that for a while, but once I became a father myself, I thought back on those moments. His callousness and disregard pained me. I never understood how someone could forget their own child. It showed me just how easy it can be to pass down cycles of pain when we don't stop to confront them.

The trauma I inherited wasn't just a burden—it was a challenge, a test of whether I would continue the cycle or break free from it. Understanding my family's history didn't erase the scars or lighten the weight. Instead, it taught me something invaluable: *awareness is the first step toward healing.*

Generational trauma doesn't define who we are—it shapes where we start. It's what we choose to do with that pain—whether to carry it forward or to let it go—that determines who we become.

CHAPTER 4
BROTHERHOOD

By the time my siblings and I arrived in the world, whatever charm Dad used to win Mom over had vanished. Outwardly, he could always put on a pleasant face and a smile. He'd hold doors for women or share a kind word with the grocery clerk, but at home, away from the eyes of the world, we knew a very different man.

To put it simply, he wasn't the kind of person you wanted to be around. If he wasn't being an outright asshole, he was depressed. He was angry at the world, and more specifically, at us, seemingly for simply making him a father.

You might have thought our names were "Stupid" or "Dumbass," considering how often he hurled insults at us. When he came home from work, he would walk through the door with a scowl, complain about his day, sit silently in his recliner for the rest of the night, or find an excuse to criticize or beat one of us. I can count on one hand the number of times he asked, "How was your day?" or said, "Tell me about school." If he did, it was usually just an introduction to get us to help him with a chore.

Kindness just seemed to be too much effort for him. In short, my father was a cruel man, and he was good at it.

Cruelty was second nature to my father, but at least I wasn't alone in enduring it. I had Mike. We arrived in this world together; I was born first, and Mike followed forty-five minutes later. Even as kids, it felt like we shared more than a birthday—we shared an unspoken understanding of what it meant to survive in our house.

In every childhood memory, my brother is right there. Together, we navigated the chaos of our home, finding refuge in each other when there was none to be found. Life in our house wasn't easy, but it taught us that survival was easier with someone who had your back. Our lives existed in unspoken parallel, inhabiting our own world within the four walls of our shared bedroom. We built a sanctuary that could withstand the tempers, the insults, and the uncertainty that greeted us outside that door. It was our safe place, our haven away from the chaos of our father.

Our room was small, about one hundred square feet, with bunkbeds and plenty of animals. We kept a reptile cage on top of our dresser and one on a small desk under the window that housed a rotating cast of wildlife. Mike and I would save up our money and go to Strictly Reptiles, an exotic pet store about thirty minutes from our house and buy whatever we could afford. We were allowed to have any reptiles we wanted as long as we cared for them and Mom couldn't smell any odors escaping from under our bedroom door.

One of our favorite pastimes was catching wild animals. Living in the Everglades meant there was no shortage of toads, snakes, and lizards. We would wander around our neighborhood with a pillowcase, lifting old boards or rocks in search of snakes. Once we caught one, we would stuff it into the pillowcase and take it home. We kept everything we caught. We had so many that we had to ride our bikes to the local pet shop weekly to buy mice for our growing reptile collection.

Once, we even convinced Mom to let us get an eight-foot reticulated python. We had to build a cage in the backyard because it was too big for our room. To this day, I have nightmares about turning the handle on our homemade wooden latch and crawling into the cage on all fours to retrieve the water bowl. Mike and I took turns with this chore, or at least that was the original agreement. The truth is, I hated going in there, and Mike couldn't care less. That is, until one day, he crawled in there, and the snake started sizing him up to eat.

With nowhere to go, Mike was cornered, and the snake knew it. That python watched Mike as he coiled his eight-foot body into a striking position, and before I could scream, "Watch out!" he launched towards Mike, biting down on his right arm. He bit Mike so hard it took both of us to pry him off. Once we washed all the blood off, we noticed the snake's teeth were still in his arm, and I had to pull them out one by one!

But nothing could deter Mike from his love of animals. He constantly read about them and had a collection of National Geographic cards that described every reptile in the world. He studied those cards tirelessly, recalling facts from memory. He would even beg the science teachers to let him keep the program's reptiles and tarantulas over the summer break.

Despite our bond, our differences were hard to ignore. For one, Mike was always the more attractive twin. He was tall, slim, and tan, while I was short, overweight, and pale. Mom made the same meals for us, but somehow, I ballooned in size while Mike stayed lean as a rail. This made me incredibly self-conscious. Whenever we swam at the pool, I felt ashamed to take my shirt off next to Mike, so I would just swim with my shirt on. The girls always had crushes on him and hardly noticed me. In addition to being overweight, I had buck teeth and a giant overbite, which made smiling humiliating as my teeth fanned out awkwardly.

I didn't know how to deal with my insecurities, and sometimes, I took them out on Mike. Since I was heavier than he was, I could win in a physical fight, but Mike had another way to hurt me: with words. One night, as we sat down for dinner, Mike looked up at me and said, "You know, John, you look just like the fat kid in that book, Blubber." Blubber is a children's novel about a 5th grader being bullied for her weight; it was eventually banned in 1993 for its lack of moral tone[i]. Immediately, I jumped up from my seat and proceeded to beat him up. But his words hurt, probably more than I hurt him.

Still, despite our differences, Mike was my best friend. At the end of the day, we had each other's backs and counted on one another. I'll never forget the night he took one of Dad's beatings for something I did. The details of the incident aren't what matter—it's the fact that, without hesitation, he stepped up for me.

We had a rubber mat in the bathtub meant to keep us from slipping. It was old and grimy, but for some reason, we loved peeling it up and sticking it to the wall. Something about the little suction cups on the bottom fascinated us, though it drove Mom crazy. She must have told us a dozen times to stop, but like most kids, we didn't listen. One night, after our bath, I pulled the wet mat off the tub, stuck it to the wall, and forgot to put it back. Not long after, our bedroom door flew open. "Which one of you did it this time?" Mom demanded.

I was already in trouble for something earlier that week, and Mike knew it. Without hesitation, he stepped in. "It was me, Mom," he said. When Dad got home, as promised, he took the beating. I never forgot that night. Mike didn't just protect me—he showed me what loyalty and love truly looked like.

It was us against the world, and in the chaos of our home, my brother and I became more than just siblings—we became each other's lifelines. Every fight, every laugh, and every act of quiet loyalty forged a bond that chaos couldn't break.

Even when we fought each other, we were still fighting for the same thing: a sense of safety, a piece of peace. Looking back, I realize that the bond we built during those years was not just about survival—it was about learning how to love, protect, and forgive. That bond didn't just help us survive; it shaped who I am today.

CHAPTER 5

THE DEPTH OF OUR ROOTS

Our roots shape us, connecting us to the strength and stories of those who came before. While it wasn't always easy to see the good in the stories of my family's past, the deeper I looked, the more I appreciated the foundation they provided me with. The same roots that carried our trauma also carried our ancestors' lessons, demonstrating that resilience doesn't grow despite struggle but *because* of it. In their struggles, I found a road map—an imperfect guide to enduring challenges and finding purpose in survival.

Our grandparents, Pat and Clara, lived in Miami, Dade County, about forty minutes from our house. We spent many weekends of our childhood with them. Grandpa was tall and good-looking, his skin tanned from the Florida sun and his Italian roots. He had strong hands and a sturdy, muscular build from many years of manual labor. His face was smooth and narrow, with only wrinkles in the corners of his mouth from his upturned, gentle smile. He was soft-spoken with a quiet but contagious laugh.

Grandma was the opposite of Grandpa in nearly every way. She was short with dark, curly hair and a busty chest she liked to flaunt in the one-piece bathing suit she wore around the house after she began experiencing hot flashes. She was a firecracker of a woman, a true Italian: loud and always making her presence known.

Their small house was in a poorer section of the city, with two bedrooms and two bathrooms. The neighborhood was full of other immigrant families just like them. When you walked outside, you could smell the rich aroma of spices drifting from the windows of nearby homes, many belonging to Cuban and Haitian migrants fleeing corrupt governments. Tall palm trees lined the streets, set against a backdrop of graffitied bridges and concrete walls.

Life wasn't easy in their neighborhood. Gang violence often erupted between rival communities, their presence summoned by the sound of gunshots echoing through the neighborhood, waking us at night. Once, Grandpa awoke to the noise of rustling outside his window only to discover someone trying to steal his car. Yet, despite the chaos that seemed to exist just outside their door, I always felt a sense of safety in their home that extended well beyond what occurred on the other side of those walls. There was a continuity, a rhythm of being with them. I always knew what to expect, which wasn't always the case in my own house.

I spent many afternoons with Grandpa in his garden. He had been a farmer in Italy and knew how to grow fresh vegetables. The smell of basil still takes me back to the days of picking it off the stem and letting its scent flood my senses. He grew fresh tomatoes, lemons, figs, eggplant, zucchini, and peppers—you name it, he grew it all. When we finished picking, we would carry everything back to the kitchen, where Grandpa would teach me how to cook authentic Italian food. He'd put on a classic Italian opera record and speak the instructions to me in Italian, saying, "Fai attenzione (pay attention)!" or "Hand me that mappina (dish towel)!"

He cooked slowly and deliberately, covering the entire table with flour before placing the dough on top and rolling it out methodically to make pasta. Everything he made was from scratch, created with his own two hands. There was always fresh bread, never store-bought, and sometimes the whole house would fill with the aroma of rising dough. Those afternoons taught me more than how to cook—they taught me patience, dedication, and the value of creating something from nothing.

Grandma worked at the county courthouse, and when she burst through the door in the evening, her voice could be heard from ten blocks down the street. Even her normal talking voice sounded like a yell. That woman wouldn't know how to whisper if her life depended on it. Over time, Grandpa learned to develop selective hearing. She'd be yelling his name or cursing from the other room, and we would have to tap him and say, "Hey, Grandma's yelling at you." I can still hear her raspy, fast voice saying, "Pasquale, you son of a bitch," followed by some comment about how he couldn't read or was being lazy, which always surprised me because Grandpa was always on his feet, cooking or working in the garden. I think Grandma just knew how to dig at him in just the right ways.

In stark contrast to Grandma's intensity, Grandpa moved passively through the world. He would sit on the porch with us and share stories about coming to America and his early years as an immigrant worker. Years later, when he discovered that Mike never finished high school, he was so angry that he wouldn't speak to him for several months. "I came here to create a better life for my family," he told Mike, "I was spit on, ridiculed, and lived in poverty, all so that you could be more than I was."

Mike's failure to finish school felt like the ultimate slap in the face to Grandpa, as if he had somehow failed himself. Unfortunately, he wouldn't live to get to see Mike return to school and earn his diploma, but he would have been proud.

Grandpa was different with us than he was with Mom. Perhaps his older age had made him kinder, less aggressive, less strict, and more patient by the time we came along. I suppose time can do that to a person. It can change someone if they're willing.

Just before Grandpa died, he started begging Grandma to call my mom, but he never gave a reason. He'd just say, "Call Patty Ann, call Patty Ann." Eventually, she got fed up and put the phone up to his ear one day and said, "Here. Now tell her what you need to say." My mom wasn't on the other end, but he didn't know that.

My grandmother never told my mom what he said on the phone that day. But in my mind, he was calling to apologize for knowing what happened to her as a kid and doing nothing to stop it. At least, that's the conversation I want to believe. Maybe it's my way of thinking that repentance and reconciliation are possible in my family and that my grandfather was somehow atoning for the long cycle of shitty dads of the past. Even if I never know the truth, that possibility gives me hope—that change, no matter how delayed, is always possible.

My grandfather came to America with nothing but hope and determination to build a better life. He faced relentless hardships—poverty, prejudice, and back-breaking labor—but he never gave up. He planted the seeds of perseverance in our family, and those roots give me strength today. They remind me that even in the most challenging times, there's an opportunity to rise above, adapt, and create something better. Strength is our inheritance, a legacy that shows we can overcome and thrive no matter where we start.

CHAPTER 6

THE SILVER LINING

One of the many spoken realities in our home was that we were poor. We were constantly told, “Money is tight" and “We can’t afford that.” However, my parents didn’t need to say anything. It was a feeling that a kid recognized the more they observed their life compared to the world around them.

Even as kids, we had to work around the neighborhood during the summers to earn a few extra bucks. Our neighbor owned a sodding business, and he paid us six dollars an hour to lay sod at homes around town. It was truly awful work. We carried heavy pieces of sod in the South Florida summer heat and humidity while spiders and fire ants crawled up our legs all day. I would get welts and bites so bad it hurt to stand up straight. Yet, we needed the money.

The same year we started at the sod company, our boss had bought his kids new bikes for Christmas and “donated” their old ones to Mom and Dad. They slapped big red bows on them and put them under the tree for us. I woke up that morning ecstatic, and I remember the look on Mom and Dad’s faces, beaming with pride that they were able to give us the bikes we had been wanting for years.

Once we finished unwrapping presents, we took our new bikes out into the neighborhood, eager to show them off. A few streets away, we came across the boss's kids. They were also showing off their new sets of wheels. As we got closer, one of them pointed at me and shouted, "Hey, that's my old bike!" My face turned red with embarrassment. I pedaled away as fast as I could and never rode my bike down their street again.

It would have crushed my parents to think I was embarrassed by our financial situation, so I never spoke of it, but I quietly took note. They were doing their best, and I was certain of that. We didn't always make it easy on them, either. Mom was nurturing and kind in ways that Dad wasn't, but she was also strict. Groundings were commonplace in our household.

We would get grounded for talking back, fighting, not doing our chores, and always for having a bad school report card. When that happened, we stayed grounded until the next report card came out and showed improvement. Being grounded in our household meant we could not leave the house, and there was to be no entertainment of any kind.

One afternoon, Mike and I came home from middle school to find Mom waiting in the kitchen, eager to see our report cards. We both hoped she had forgotten, but when she asked us to open our backpacks, we knew what was coming. We each managed to receive poor grades that quarter. We pulled our report cards out and set them on the table. As Mom picked them up and began scanning the grades, a look of disappointment crossed her face, quickly replaced by anger.

We knew the drill. Mom sent us both to our room, saying, "Don't even think about coming out." We walked back, ready to settle in, and braced ourselves for the hell that was sure to rain down on us as soon as Dad got home.

An hour later, we heard the front door open, and heavy footsteps headed towards the kitchen, followed by muffled voices. The next thing we knew, the lights in our room suddenly went dark. Dad had shut off the power so that we couldn't play our new Nintendo game, which we had just rented. We were left in a dark room with nothing to entertain ourselves but the sound of our own voices.

It was Christmastime, but it was warm in Florida, so Mike and I decided to open our window. As I cranked the handle and lifted the glass, I had an idea. Our bedroom window faced the front of our house, and Dad had recently strung Christmas lights around the hedges below it. My eyes followed the end of the light trail until they landed on an extension cord. I turned to Mike and grinned.

Mike and I pushed the screen out of the window, and I dove out headfirst while he held my legs. I grabbed the string of lights, pulled the extension cord into our bedroom, replaced the screen, and closed the window. We plugged in the television, muted the volume, and played Nintendo all night. When Dad opened our door and caught us red-handed, he delivered our punishments just as we had expected, but still, we felt like we had won that day.

Fortunately for us, Dad wasn't home much. He was a firefighter who worked twenty-four-hour shifts several days a month, leaving Mom with the three of us. Dad rarely talked about his job. Sometimes, after finishing a shift, he'd come home the next morning, and instead of picking on me or Mike, he would sit quietly in his chair, staring off into the distance. I never knew what kinds of things he had witnessed or what sort of terror or loss of human life he encountered; he never said a word. When I was around seven years old, he fell through the roof of a church while on the job, and when he landed, he fractured his back and went on permanent disability. He never returned to work with the fire department.

But on the nights Dad was away, Mom always made us breakfast for dinner, which we loved. One night, it might be waffles, and the next, eggs and pancakes. These dinners always felt like a small celebration, though none of us would dare call it that. You could sense the air in the room lightened when Dad wasn't home, as if a heavy fog had lifted, allowing us to breathe, even if just for one night.

I can count on one hand the number of times my father said, "I love you," and even fewer times when I heard him say, "I'm proud of you." When Dad was home, I felt as if I were walking a tightrope along a razor's edge—one misstep could send him into a rage. We never quite knew what would upset him next. The only certainty was a slap across the face if Mike or I talked back or disrespected him in any way. It didn't matter if we were sitting at the dinner table with our mouths full; he would hit us so hard that our dinner would fly clear across the room and land on the wall. Yet, he was a man full of contradictions, as there seemed to be no consequence for all the times he disrespected us, our mother, or made us feel insignificant. He didn't see that as a problem.

Our parents' room was next to ours, and often, we could hear them shouting at each other from the other side of our wall. One night, we were in our beds when we heard a commotion coming from the living room. The shouting grew louder as they moved down the hall toward their bedroom. There was a loud slam as Mom shut and locked the bedroom door before Dad could follow her in.

The next thing we heard was Dad screaming, "You better open this door, or I will break it down." Mom stood her ground; she didn't open the door. Then, we heard a thud followed by a crash. Dad had broken down the door. He had too much size on both of us for us to physically do anything to help Mom. Instead, we peeked from beneath our covers through our open bedroom door, terrified and helpless, pretending to be asleep, afraid that his rage might spill over onto us.

We cried in our beds all night, and the next morning, we were too scared to leave our room. We knew we would have to walk past the broken door to reach the living room, where Dad would be sitting in his chair, acting as if nothing had happened.

Mom would later recount this very fight to me when I was an adult. What we kids didn't hear at the time was Mom getting inches from Dad's face and saying, "I dare you to hit me. My God, I dare you to hit me." With those words, Dad backed away. He never laid a hand on her that day or ever. Mom told me, "Him hitting me should have never been the line I drew in the sand. I should have drawn that line much earlier."

Mom made it clear that day that if he ever hit her, she would leave him, but Mike and I were a different story. I have a hard time remembering the times Dad abused me. I know it happened, but somewhere along the way, I pushed those memories into a backlog of things from childhood that I wanted to forget. However, I can remember with clarity nearly every time it happened to Mike.

One of the worst moments was when we were in eighth grade. Mike and I had just gotten home when the phone rang. It was the school calling to inform my parents that Mike had gotten in trouble earlier that day for talking back to a teacher. Dad answered the phone and said to the administrator on the other end, "He did what?" followed by, "Well, rest assured, I will handle this."

Mike was in the shower, so he had no clue what was coming for him. Dad waited outside the bathroom, and as soon as Mike opened the door, with his towel still wrapped around his waist, Dad began hitting him. Dad was punching him so hard and so fast that the towel flew off Mike's body, leaving him completely naked in the hallway while Dad continued to beat him.

I watched from the dining room, feeling powerless to help him. The fear was as palpable as if I were the one taking the beating. When I was on the receiving end of Dad's blows, I could dissociate and take my mind elsewhere, but when he targeted someone I loved the coping mechanisms that had helped me through my own fear of being beaten disappeared. All that remained was a sense of helplessness. I would have rather taken the beating myself than witness the humiliation and degradation of my twin brother, my other half.

When Dad got angry, everything about him changed—his voice would suddenly drop to a lower octave, deep and gruff, like a villain in a superhero movie. His eyes turned black, cold, and empty as if he were looking right through you. I learned to pay attention to even the smallest change in Dad's demeanor or body stance, the shift in his gait, or the specific way he would move his hands. I studied my father the way other kids might study the ecosystem or biology, understanding how even the slightest change could disrupt the whole environment in which the organism or human thrives and survives.

My mind and body learned to recognize these threats of danger. I developed such a deep understanding that I could sense a change in someone's heart rate simply by being near them. I would later learn that these are adaptive traits of an abuse survivor. They are the tools that help survivors cope and observe impending threats, even if they feel powerless to prevent the harm itself. But I didn't have the language to call it abuse or to understand that what was happening to my brother and me was not just physical; it was psychological, too.

In the 1980s, The National Committee for Prevention of Child Abuse began running public safety campaigns to raise awareness of child abuse occurring across the country[ii]. Harsher penalties were enacted for parents accused of abuse. Resources became easily accessible for children to report instances of domestic abuse happening in their homes.

One of those commercials aired while we watched television after dinner one night. It featured a series of actors portraying various parents, zooming in on their mouths as they repeated phrases like, "You're pathetic; you can't do anything right," "Why don't you go find somewhere else to live?" and "I wish you were never born," all directed at a child on the receiving end of their words. These are all things Dad would say to us. At the end of the commercial, a hotline number was displayed. Dad looked at the three of us and said, "None of you better ever call that number because if you do, they'll haul me off to jail. I can promise you; it'll be a hell of a lot worse for you once I get out."

But like most things in life, not everything about Dad was bad. There were moments when he tried to show us that he cared. On weekends, Mike and I would spend hours outside with him, fixing up the house, working on cars, landscaping, or building things. He taught us how to do practically anything ourselves or how to find the books and resources to figure it out.

Once, we bought an old, beat-up fishing boat from our neighbor, and the three of us spent months restoring the hull and rebuilding the engine.

When we took it out on the water for the first time, Dad might not have had the right words, but if there was ever a moment of joy I saw in him, that day was it. What Dad couldn't provide in affection, he made up for with practicality and skills. Perhaps that was the only way my old man knew how to show love.

Truth be told, I come from a family filled with child molesters, lousy dads, deserters, and abusers. For the longest time, I overlooked my parents' shortcomings because I thought they did the best they could with the hands they were dealt. Justifying how they treated us was due to their own fucked up childhoods.

Until I became a father myself. Now I have the wisdom from my own experiences, and I learned you don't have to be defined by the mistakes of those who came before you. You have the power to choose a different path, break the cycle of negativity, and be a positive influence, especially as a parent. When I became a father, nothing else in the world mattered more than making sure I showed up every day for that little girl. Despite my troubled family background, I was determined to be a better man for her and end the cycle of bad dads.

But breaking the cycle didn't mean erasing the past—it meant learning from it. The very instincts I developed as a child, the ones that kept me safe in an unpredictable home, didn't just disappear. Instead, they sharpened into something else entirely. What began as an adaptive trait for survival through years of abuse has now evolved into an intuitive skill. The ability to perceive subtle changes in the room, down to someone's heart rate increasing, has become a talent that I have refined and utilized to my advantage. It has enabled me to thrive in all areas of my life, particularly in the workplace. This form of emotional intelligence, developed from my childhood trauma, is extraordinary and cannot be taught in school. In a strange way, I am grateful for all the years my father abused me because, unbeknownst to him, day after day, year after year, he was helping me refine my craft. It's my *silver lining*.

CHAPTER 7

A "HIGH" PRICE TO PAY

The first time I tried drugs; it didn't feel like a life-changing moment. It felt harmless—a fleeting decision born out of curiosity. It was just a "hit," nothing more, or so I told myself. However, looking back, that single choice came with a hidden cost I couldn't yet understand. What seemed like a small step at the time was actually the first stride down a path that would demand far more than I was willing—or ready—to pay. That first "high" offered a glimpse of escape, but opened a door I wouldn't know how to close.

At thirteen, I had no idea something as small as a single hit could ripple through my life with such devastating consequences. It started with weed—the so-called gateway drug. I never felt pressured to try it; rather, it was more of a pull, a quiet allure that came from watching a friend. That friend was a classmate from seventh grade who would get high at school while I simply watched, having no desire to get high myself.

But even then, I noticed a seed of curiosity being planted, one that would eventually take root. Our friendship continued through middle school and eventually evolved into him coming over to my house after school or on weekends. He would pick me up on his four-wheeler, and we would head straight to the trails and ride through the Everglades.

While riding, we would eventually find a place to stop. I would light a cigarette while he smoked weed. Initially, there was no peer pressure or shame to get me to smoke with him; neither of us cared. We were just interested in exploring and breaking a few rules along the way.

Then, sometime in eighth grade, during one of our usual weekend adventures exploring a new part of the Everglades, I watched him pack that old glass bowl full of weed—and something shifted inside me. A sudden wave of curiosity hit me, stronger than I expected. The scent of the weed as he prepared it was rich and earthy, strangely enticing, pulling me in. Without thinking, I asked if I could take a hit. His eyes lit up with excitement. "YES!" he shouted, passing it over without hesitation as he explained to me how to “hit” the pipe.

Smoking, in general, was not new to me; I had been exposed to it my entire life. Dad smoked cigarettes for as long as I could remember, but Mom never let him smoke in the house. So, he would go outside to smoke in his truck. Sometimes, after dinner, Mike and I would sneak out to Dad’s truck and take a few of his cigarettes. If there weren’t enough in the pack to grab without him noticing, we would pick up some of the burned cigarette butts he had thrown out on his way back into the house. We would take our fresh stash of cigarettes, climb into the treehouse in our backyard, and smoke them as if we were the coolest kids on the block.

So, when I took that first hit of weed, I expected it to be just like smoking cigarettes. However, the moment I took a puff and inhaled, my lungs felt like they were on fire. I coughed so much I thought my lungs would detach from my body and land on the ground in front of me.

Even though I cannot recall what that first high felt like or many details of the experience, I do remember that when I returned home and walked inside my house, no one noticed. No one noticed I was high; no one noticed my bloodshot eyes; no one asked any questions when I raided the refrigerator. I had just smoked an illegal substance and walked away scot-free. I fooled everyone, including myself. Proving every parent and teacher wrong who insisted that doing drugs would kill me.

In the 1990s, the D.A.R.E. (Drug Abuse Resistance Education) program was in full operation in the Florida school systems[iii]. This program allowed schools to show pictures and videos of people using drugs and the consequences that followed. These typically featured horror stories and graphic images of individuals suffering from various addictions. It served as a scare tactic aimed at educating students on the potential outcomes of trying drugs. Yet, for me, it only reinforced the illusion that I could outsmart the system. I had just smoked weed, and nothing negative had happened. So, I tried it again. And then again, until it became a regular fixture in my social life. In the end, "the gateway drug" lived up to its name, as it truly marked the beginning of my drug journey. What seemed so innocent back then would ultimately derail my life in more ways than one.

Once I entered the social circle of drug users, I discovered there were numerous options available to me. Some of my friends were into stimulants like cocaine and speed, while others preferred depressants like Roofies and Xanax, and others enjoyed psychedelics. I, however, was different—I liked it all.

Within a few months of smoking weed, my drug use escalated quickly to acid and cocaine. John, another friend, lived only a few streets away, so I would ride my bike to his house to hang out and smoke. John was older and had a car—a 1972 Barracuda. He came from a well-off family; his dad was a high-ranking police officer and a car enthusiast, while his mom owned a local Dairy Queen where I worked one summer.

One day, John was waiting for me as I pulled up on my bike. He was sitting inside his vintage hot rod, warming it up. He said, "Hop in. I need more weed." When we ran out of weed, we would meet up with his buddy. This guy was the spitting image of Matthew McConaughey in Dazed and Confused, with his straight blonde hair and stubbly mustache. But this time, he was offering more than just a bag of weed: "You wanna do some acid, man?" he asked as he brushed his hair back behind his ear.

I didn't hesitate when he handed me a tiny strip of paper with a peace sign stamped on it. I held out my tongue and pressed the paper down until I felt it dissolve into oblivion. The next thing I remember, I was standing in acid man's gravel driveway, staring down at the rocks. I methodically picked them up, only to watch in slow motion as they fell back down to my feet. At some point that night, I must have thought it was a good idea to buy a gram of cocaine from McConaughey's doppelganger because when I woke up the next morning, I found a gram of cocaine in my pocket. So, I did what any good friend would do – I called my buddies over, and we snorted the entire bag.

Once I started dabbling in drugs, I quickly realized that nothing was off-limits. I always seemed to know someone who had access to something—some new high, some forbidden thrill. At first, it was just a social thing, something I did with friends, a way to have fun and escape. There were no immediate consequences, nothing to signal the danger ahead.

What I didn't see—what I couldn't see—was that I was already on a downward spiral, picking up speed with every choice I made. The path I was on had only one destination, and the only way out would be a devastating crash. That first "high" was never truly free. It came at a cost I wouldn't fully understand until much later, when I realized the price was not just the drugs themselves, but also the pieces of myself that I gave away in pursuit of them. The sense of control I believed I had faded with every decision that followed. Addiction doesn't ask for permission; it takes, piece by piece, until you're left wondering how it all began.

And for me, it began here—with one seemingly innocent choice that forever changed my life. A choice that proved every high comes with a price.

CHAPTER 8

THE MOVE

We sold our house after my sophomore year of high school in 1997. My parents had tried to sell it several times before, but Hurricane Andrew had other plans. When the storm tore through south Florida in 1992, it left a path of destruction that reshaped entire communities. To this day, Andrew remains one of the most devastating hurricanes in Florida's history, and until Hurricane Katrina in 2005, it was the costliest storm ever to hit the U.S. The damage went beyond buildings, and our neighborhood, like so many others, never fully recovered.

After the storm, my parents took the house off the market, waiting for the dust to settle and property values to stabilize. But by the time my sister graduated high school in 1997, they were ready to try again. But this time, the house sold, and instead of buying another home in Florida, Dad made the decision to move our family to Sevierville, Tennessee. For me, it wasn't just a decision—it was a blindside. One day, Florida was home, and the next, it wasn't.

As kids, we had taken a few family vacations to the Smoky Mountains, spending a week in a cabin in Tennessee or North Carolina. But I could count on one hand the number of times Dad had joined us on vacations; most trips were just Mom and a few of her friends with kids our age. Despite having little connection to Tennessee—no real experience there, no family nearby—Dad somehow decided that this would be our fresh start, uprooting our lives with little more than a vague plan and a change of scenery. Years later, Mom described the move as part of Dad's "insatiable need to pursue the next thing that might finally bring him happiness." She knew deep down that no change of scenery could cure his depression, but if there was even a slight chance, she was willing to take the risk.

But Mom had her own coping mechanisms. Over the years, she developed her own vices—first food, then shopping, followed by an unchecked need to spend. Looking back, I see now that these weren't just bad habits; they were survival mechanisms. Like the rest of us, she was just trying to endure life with my father. Having come from her own troubled background, she had never been shown how to process emotions in a healthy way or how to withstand the weight of his unhappiness.

Only later in life do you begin to see your parents' struggles for what they truly were. As a child, your family life is simply *normal*—the framework through which you understand how the world works. It's not until you step outside of it, gaining distance and perspective, that you start to recognize the cracks, the patterns, and the weight of what you once accepted as just the way things were.

That shift in perspective hit me hard when we moved to Sevier County, Tennessee—the hometown of Dolly Parton, tucked in the shadows of the Great Smoky Mountains. In the early 1900s, it was little more than a patchwork of subsistence farms, but the opening of the Great Smoky Mountain National Park in 1934 transformed it into a tourist destination. Visitors came in droves, seeking a brief escape into nature. But while they came to *get away*, we had arrived to *stay*—and for me, settling into this new life felt anything but ideal. Our new house was a cramped rental, even smaller than our home in Florida. That summer was stifling, not just from the heat but also from the overwhelming sense of displacement. Mike and I relied only on each other as we tried to navigate a world so different from what we had known.

Florida was where we left our roots, our friends, and the carefree days of childhood. Tennessee was where we would need to start over, redefining home, family, and the parts of ourselves we had yet to understand. It was a strange and lonely beginning, but even then, we knew it wasn't an end—it was the start of something entirely new.

CHAPTER 9

A DANGEROUS GAME

Predictably, Dad wasn't any happier in Tennessee than he had been in Florida. Mom had hoped that maybe, just maybe, the move would give him a fresh start and bring some much-needed peace to our family. He'd promised her as much when he sold her on the idea. Instead, nothing changed—he carried his misery with him, and it weighed just as heavily on the rest of us as it always had.

When summer ended, it was time to start our junior year at our new high school. We felt relieved to have a place to go that wasn't just our home, even if it was school. Sevier County High School, with only twelve-hundred students, felt small compared to our Florida high school, which had over six thousand. The hallways were quieter, the faces more familiar, and the small-town atmosphere was a stark contrast to what we had known.

Despite our initial resistance to the move, Mike and I eventually made new friends, and our bitterness about leaving Florida began to fade—at least a little. But old habits die hard, and the crowd we gravitated toward looked a lot like the one we had left behind: the kids who liked to party. If we were searching for a fresh start, we weren't exactly heading in the right direction. At school, academics took a backseat to more pressing concerns—what girls thought of me and what was happening on Friday night. I wasn't exactly popular, but I wasn't an outsider either. I floated between different groups, with alcohol and parties serving as my main social currency. Give me a few beers, and I could turn into the life of any party.

But some nights had a way of cutting through the haze, reminding me that we weren't as invincible as we liked to believe. One afternoon after class, we overheard friends whispering about a party—a common ritual in the days before cell phones. Weekend plans weren't texted or posted; they spread through hallway chatter and word of mouth. "Hey, there's a party tonight. Meet at 8 o'clock at Patriot Park," someone said. Without GPS or instant messaging, meeting up required coordination. Bigger parties often had staggered arrival times—one group at eight o'clock, another at nine, then ten—to make sure no one got lost or left behind.

That night, a few friends and I met at the park and joined the line of cars heading to the party. We were near the front, leading the way down dark, winding roads. As we arrived, we pulled off and waited for the rest of the caravan to catch up, including a car carrying one of my friends.

But the next vehicle to appear wasn't theirs. It was a patrol car, blue lights flashing. Most of my friends scattered at the sight of the cops, but a few of us stayed put. The officers told us there had been an accident. Someone had died. They wouldn't say who. We had been drinking, and they knew it, but instead of arresting us, they told us to stay put. If they saw us drive off, they promised we'd end up in jail.

So, we stayed. That night, we slept in our cars, surrounded by darkness, the flashing red and blue of emergency lights still burning in the distance. We didn't know who was gone, only that life could change in an instant—and that none of us were untouchable.

It wasn't until the next day that we learned it was someone from our school. He was a friend of mine who sat next to me in class and made everyone laugh. Returning to school on Monday and seeing his empty chair felt different.

That weekend changed everything. For weeks after the accident, the weight of his loss lingered heavily in the halls of our school. Whispers and rumors surrounded the devastating wreck and the events leading up to it, as if everyone was trying to piece together how something so tragic could happen. The parties slowed down, replaced by a collective sense of fear and vulnerability that hung over those of us who had been there that night. For the first time, it felt real—how quickly everything could be taken away. It could have just as easily been any one of us who lost our lives that evening.

In the years that followed, his funeral would be the first of many. One by one, we said goodbye to friends we had grown up with, and over time, it became a tradition to bring a shovel to the graveside service. After the casket was lowered into the ground, we would take turns shoveling dirt, one scoop at a time, until our friend was buried. To this day, when one of our core group of friends passes away, we all come together, shovels in hand, to honor the tradition we promised each other so many years ago.

Despite those moments of grief and loss, life moved on in ways that felt both normal and destructive. Drug use had become a constant presence in my life, weaving itself into both my daily routine and my social connections. It started innocently enough with smoking weed before school, but it wasn't long before I'd also secured a steady supply of hydrocodone painkillers through a friend. At first, it felt like a harmless hustle—taking pills to school, selling a few to friends, and keeping the rest for myself. But it didn't take long for that hustle to spiral into something more.

Our school had a resource officer on site, and eventually, word got around that I was selling pills. Remember when I mentioned having a rotating cast of friends who moved in different circles? Well, one of those circles included the smart friends—the ones who knew how to stay out of trouble. And one of those smart friends volunteered in the front office. What can I say? It pays to have intelligent friends. This particular friend overheard my name come up in a conversation between the administration and the officer, and without hesitation, he slipped out of the room and bolted straight toward me.

I was at my locker during the class change when I spotted him sprinting through the crowded hallway. His expression said it all before he even spoke. "John," he gasped, "They're looking for you. They know about your stash. You've got to get rid of that shit—NOW."

Thankfully, the hallway was packed with students changing classes, buying me some precious time. I realized the resource officer would need a while to push through the crowd and find me, but I also knew my options were limited. I had to get rid of the pills—fast. Flushing them seemed like the safest bet, but the nearest bathroom was right by the school office, and heading in that direction felt too risky. Instead, I quickly gathered a group of my friends around me, pulled the hydros from my pocket, and said, "Everyone needs to take one." No questions asked—they each grabbed a pill and swallowed it.

That left me with four pills. There wasn't enough time to do anything else, so I tipped my head back and swallowed all four in one motion. Until that moment, 20mg was the most I had taken at once. I wasn't about to go to jail for possessing narcotics, and if 20mg made me feel great, I figured 40mg was going to be one hell of a ride.

I walked into my next class, and sure enough, a few minutes later, the officer and the principal showed up at the door: "John Mathews, you need to come with us," they said. By that time, the hydros had fully kicked in, and I could feel the buzz rushing through me. Trying my best to act normal, I followed them down the hallway to the principal's office.

"Empty your pockets," the officer demanded, watching me closely. I reached into my pants and pulled out a few empty gum wrappers and some quarters. He looked stunned—and frustrated. After finding nothing, they marched me down the hall to my locker, forcing me to empty every shelf and notebook inside. Once they finally accepted they weren't going to find any pills, they let me go.

In that moment, all I could think about was how I had outsmarted the system. I felt untouchable, convinced I was one step ahead of everyone. What I failed to see was just how close I'd come to a life-altering disaster. If they had caught me that morning, there's no doubt I would have been arrested and expelled. The consequences would have been devastating, derailing my life before it had even really started. But at the time, none of that crossed my mind—I was too busy celebrating my "victory."

That wouldn't be the last time I had a run-in with authority. A few months later, while sitting in class, I was interrupted yet again. This time, the police walked in without so much as a knock. They didn't bother with polite requests. They pointed straight at me and said, "*You*, come here." I was just as stunned as the rest of the class. My mind raced as I tried to replay everything I had done that morning, but nothing made sense. As I followed them outside, my confusion turned to panic when I saw police cars surrounding my brother, Mike's truck in the parking lot. Frantically scanning the scene, I spotted Mike standing beside his vehicle—already in handcuffs.

Still unsure of what had happened, I walked closer and noticed a K-9 dog inside his truck, sniffing around the cab. Apparently, the school administration called the police and asked them to bring a drug dog. Once the dog was released, it headed straight for Mike's car. Luckily for both of us, there weren't any drugs in the truck, or on us—it just reeked of weed from the night before. Once again, we walked away unscathed. But instead of feeling relieved, a small voice in the back of my mind whispered to me that our luck might eventually run out.

But for every reckless decision I made, it seemed fate was right there, waiting for a chance to intervene—sometimes offering a lesson and other times presenting a gift I wasn't ready to understand. That summer before my senior year, fate arrived again, this time in the form of a beautiful girl named Mandy Hildreth.

CHAPTER 10

SUMMER FATE

The summer before my senior year, fate came disguised as an ordinary evening when my buddy pulled up in his car. I wasn't expecting anything life changing. But then, I saw *her*.

She was sitting in the front seat with her feet propped up on the dash, wearing a casual t-shirt and shorts that hinted at a carefree summer spirit. Her dark, curly hair framed her face in a way that made it impossible to look away.

The moment I laid eyes on her, the world slowed as if the very air shifted to mark it. I climbed into the back seat, fighting to play it cool. "I'm John," I said, introducing myself.

She turned around, and when she smiled, it felt like the earth tilted on its axis. Her eyes sparkled, carrying the warmth and mischief of someone who could break a heart without even trying. "I'm Mandy," she said. Her voice was soft yet confident, and the way her lips curved into a grin made me smile back, too—like I didn't have a choice in the matter.

From that moment on, I was hooked. I hung on to every word she said, laughing when she laughed and doing everything

I could to earn more of that smile. But just as quickly as the connection sparked, the reality of her situation set in: Mandy was dating a guy named Nick. And I knew Nick—I had been "talking" to his sister, which in high school terms was the step leading up to dating.

Still, there was something about the way she looked at me that lingered. Something unspoken, like maybe the universe had brought us together for a reason.

Fate played its next hand when I landed a summer job at Adventure Raceway. Mandy worked the ticket booth, giving me a reason to be there more often than I should. I didn't confess my feelings; I didn't need to. Just seeing her made every shift worthwhile.

When I heard she and Nick had broken up, it was like the floodgates opened. My relationship with Nick's sister had fizzled out the moment Mandy entered the picture, and now that she was single, I didn't waste a second longer. I pursued her relentlessly, finding every excuse to talk to her and trying to charm my way into her world.

She didn't make it easy. Mandy wasn't one to fall for smooth talk or desperation, so she kept me at arm's length. However, even though she didn't agree to a date that summer, our connection was undeniable. We spent time together—sometimes as friends, sometimes as something more—but I never lost sight of what I wanted.

There were moments when we drifted apart, times when we dated other people. But even then, Mandy remained at the center of my thoughts. I couldn't shake the feeling that she wasn't just a girl I wanted to be with—she was *the* girl. The one who had unknowingly taken up residence in my heart the moment she turned around and smiled at me from the front seat of that car. Somewhere, deep down, I knew this story between us was far from over.

CHAPTER 11

PAY IT FORWARD

By the time my senior year rolled around, I had completely given up on school. When your father has called you a dumbass since you were old enough to read, it's hard to believe you're capable of anything more. My grades were barely passing, and I jumped at any excuse to leave. Then, when I turned eighteen in the middle of my last semester, it gave me all the freedom I needed to stop answering to anyone. I was finally in control—or so I thought.

At first, I started leaving school earlier and earlier, testing how much I could get away with. Eventually, I stopped showing up altogether. It didn't take long for the administration to use my absences against me, reporting me to the state of Tennessee. To me, it felt like another way for them to prove they were in charge, another attempt to punish me for refusing to play by their rules. Legally, I was an adult, but under the law, parents were still responsible for their child's attendance. Once I hit the threshold for unexcused absences, the school system didn't hesitate to send me to truancy court—something they seemed all too eager to do.

On the day of my court appearance, I headed to the courthouse, joining a crowd of other so-called delinquents from across the county. One by one, we stood before the judge, each of us getting our turn in the spotlight. My twenty or so absences felt like small potatoes compared to some of the others—kids who had missed more days than they'd attended. When my name was finally called, the judge barely looked up from his papers. He gave me little more than a slap on the wrist and sent me on my way, as if to say I wasn't worth the effort.

I walked out of the courtroom, convinced I had won, my confidence swelling with the belief that once again, I was untouchable—above the rules and above reproach. The following week, in yet another act of defiance, I decided to sign myself out of school in the middle of the day, just to prove, once again, that they couldn't break me. I strolled into the office, filled out the early release form like I owned the place, and walked out without a second thought.

It didn't feel like much at the time—just another small act of rebellion in a game I believed I was winning. But in hindsight, I see it for what it was: a stone I cast into the water, beginning a ripple effect that would reach far beyond anything I could have imagined. That one decision, insignificant as it seemed, set everything else into motion.

After signing myself out of school that day, I thought I had beaten the system. I was eighteen, confident, and convinced the rules didn't apply to me. But life has a way of humbling you when you least expect it. What began as a decision to assert my independence quickly turned into a journey filled with unexpected challenges—and a lesson I would carry with me for the rest of my life.

Back then, the rules at school felt more like a game to outsmart than a system to follow. Our school was surrounded by a chain-link fence, with two gates on either side of the parking lot—one for students, monitored by a guard, and the other in the back for buses, left unattended. If you wanted to leave early, you needed a purple slip from the front office and had to show it to the guard at the main gate as proof of your permission to exit.

That afternoon, I tucked the purple slip into my pocket, pulled out my keys, and headed to my car. It was parked by the back gate, and as I sat in the driver's seat, a lazy thought crossed my mind: *Why bother with the front gate?* Without a second thought, I started the car, drove straight through the unguarded back gate, and left the school. No slip, no stop.

The following day, I was called into the office. The administration had caught me on camera bypassing the guard gate, and by that point, they'd had enough of me and my disregard for the rules. Our long-running game of cat-and-mouse had finally come to an end—they had me. I wasn't welcome back. Instead, I was sent to the Alternative Learning Center (ALC), a school reserved for students with behavioral issues.

That afternoon, the school called my father to tell him I'd been kicked out. Dad had always been clear: "If you get in trouble at school, you get in trouble at home."

When I walked through the door, he was waiting for me in the kitchen. His voice was steady as he confronted me, but I just shrugged and said, "I'm eighteen. I can do whatever I want."

While that might have been the case, I was still living under his roof. That meant his rules still applied, so he handed down a punishment and grounded me. He wasn't physically able to beat my ass anymore, so he resorted to more conventional consequences. But I had already made plans that weekend with my friends. We were heading to Ohio to visit some girls we had met at work. I spent the entire spring break gathering phone numbers from girls vacationing in Pigeon Forge, and now I was ready to cash in.

We had a hotel room booked and high hopes for the weekend. I was a teenage boy with one thing on my mind. So, when I told Dad I planned to go, despite being grounded, his response was matter-of-fact. "Look, you're eighteen, I can't stop you. But if you leave this house, when you come back, all your shit will be packed and on the porch."

I decided to roll the dice, and by Friday, I was Ohio-bound. My friends and I pooled together every dollar we had, spending it all on gas, food, drugs, and a cramped hotel room—three guys and three girls packed into two queen-sized beds. The weekend felt like a blur of reckless fun, and by the time we headed back to Tennessee, we were riding high on adrenaline and a few other things. But that high came crashing down when we realized, halfway home, we didn't have a single penny left between us to fill up the car with gas.

Desperation kicked in, and we came up with a plan to steal the gas. We thought we would be slick and leave the nozzle on the ground after pumping the gas. This way, the attendant inside wouldn't be alerted. We'd be long gone by the time another car pulled up and returned the nozzle to its proper place.

So, we parked the car at the farthest pump, and I began filling it with premium gasoline. I thought if we were going to get caught, we might as well go for the good stuff. When the tank was full, I laid the hose on the ground just as planned and drove off. We kept checking the rearview mirror for blue lights, but they never came.

When I got home, as promised, all my bags were sitting on the front porch, and the doors were locked. I pounded on the door until my dad answered, but instead of letting me in, he opened it just enough to hand me a fifty-dollar bill and, with a smirk on his face, said, "Good luck." And just like that, I was homeless.

Hearing my dad say, "Good luck," wasn't the fiery response I had prepared for. I expected anger, maybe even a lecture, but I only got two words that felt heavier than anything else he could have said. The smug grin on his face made me feel like he had been looking forward to this day.

Those two words followed me long after I left. At first, I shrugged them off, interpreting them as a dismissal, maybe even a challenge. But as the hours passed, the weight of "Good luck" began to sink in. Was he wishing me well, or was he expecting me to fail?

I loaded my two duffel bags' worth of possessions into the 1978 Oldsmobile Cutlass Supreme I had bought for two hundred dollars. I had a pack of cigarettes, a handful of joints, some leftover vodka from the weekend, and that fifty-dollar bill from my dad. With nowhere to go and no plan, I started driving. I thought I might pull the car over somewhere and sleep in the back until I figured out what to do next. In the meantime, I turned down the highway, rolled down my window, and smoked a cigarette.

After buying dinner at the local Long John Silver's, I spent the first night sleeping in the parking lot. I knew I couldn't stretch that fifty dollars for long. As I laid across the back seat of my car, covering my face with a baseball cap, hoping no one would break into my car that night, I boiled with anger. "Fuck him, fuck him, fuck him," I repeated out loud until I finally fell asleep. The following day, I climbed back into the front seat and drove myself to school.

A week later, I was still homeless. I had managed to make it out of the Long John Silver's parking lot and onto my friends' couches, bouncing from one house to another. I had no plan and no place to call home. I was still working the racetrack but barely made enough money to stay afloat.

One night, desperate to clear my head, I decided to take a drive down the road to blow off some steam. I pushed the gas pedal to the floor, and just as I picked up speed, the entire car rattled violently beneath me. At this point, it was barely even a car—it was a clunker held together by sheer willpower. The air conditioning and heat hadn't worked in ages, the radio was dead, and the gas gauge had been broken long before I ever bought the damn thing.

Then, just as I pulled onto the corner of Newport Highway and Robert Henderson Road, the engine sputtered and died. I turned the key, hoping for a miracle, but nothing happened. Frustrated, I slammed my hands against the steering wheel and yelled, "Fuck me!" before stepping out of the car. Standing on the side of the highway in the dead of night, I took a deep breath, trying to figure out what the hell I was going to do next.

Headlights zipped past me, one after another, their beams cutting through the dark as if I wasn't even there. Then, in the distance, I noticed one pair of lights begin to slow down. The car crept closer, its engine humming softly as it pulled off the road and came to a stop about twenty feet behind me. My heart raced as the driver stepped out and started walking toward me.

At that moment, I wasn't sure whether I should feel scared or relieved—grateful for the help or wary of the stranger. But standing there alone on the side of the highway, I had no choice.

When he reached me, he extended a hand and smiled. “What seems to be the trouble?” he asked, his voice calm and steady, cutting through the tension in the air.

“Best case scenario, I’m just out of gas,” I told him. “Worst case, the engine’s failed.”

He glanced at the car, nodded, and said he’d head up the road to find a gas station. About half an hour later, he returned with a full canister of fuel. I poured it into the tank and turned the key.

The engine sputtered to life—it turned out the damn thing had just run out of gas. I thanked him and instinctively reached into my pocket for the money I knew wasn’t there. Before I could say anything, he opened his wallet and handed *me* a twenty-dollar bill.

Still in shock, I asked, "What’s your phone number? I’ll pay you back.” But he shook his head and smiled. “I don’t want your money. When I was your age, I was in the same situation you are now.”

His words caught me off guard. I stood there, stunned by his kindness, but also wondering how this man in front of me could have ever been in my shoes.

He continued, “My car broke down on the side of the road just like this when I was at my lowest, and a stranger stopped to help me. Now I’m going to tell you the same thing he told me.”

I leaned in, his tone drawing me closer.

“When you make it in life,” he said, “make sure you stop and help someone else, too.”

Those words stayed with me. It took years to repay his kindness forward, but I kept my promise. Time and drugs have long erased the man’s name from my memory, but I think about him often and how much I’d love to shake his hand and thank him. So, sir, if you’re reading this book, please reach out. I would love the chance to tell you how much your kindness meant to me. You changed my life.

That day, I learned that some of the most meaningful gestures often come from strangers. You never know what someone is going through or what battles they’ve already fought, so it’s important not to judge.

This man knew I couldn’t pay him back, nor did he want me to. Instead, he asked me to pay it forward, just as someone had directed him years before. A simple concept with compounding rewards: instead of repaying a favor, you pay it forward. Whether to a stranger, a colleague, or a friend, it's like throwing a rock into a lake and watching the ripples spread in all directions—a ripple effect of generosity.

His lesson stayed with me, shaping how I view the world and how I show up for others. I’ve received so much help along the way, and because of that stranger, I’ve made it my mission to keep the ripple going, one act of kindness at a time.

CHAPTER 12

THE RIPPLE EFFECT

After several weeks of couch surfing, my friend Tony offered me a more permanent place to stay. He lived in a small one-bedroom house owned by his grandmother, and every week, she would stop by and fill our fridge with leftovers. The living room had a short couch that barely fit my legs when I tried to sleep, but it was a stable place with food to eat—an improvement from the Long John Silver's parking lot.

Every morning, I rode to high school with Tony. However, instead of going inside with him, I had to wait outside for a bus that would take me and a few other students to the ALC, which was thirty minutes away. The moment we arrived, we were reminded that this wasn't an ordinary school. Two officers greeted us at the entrance and patted us down before we passed through the metal detectors on the other side.

The "school" itself barely resembled one. Rows of desks filled a large gymnasium that had been converted into a single classroom, an open space meant to contain, not educate. Each student was assigned a desk, where we sat all day, silently working on assignments given by our individual high schools. The intention seemed to be to bore us into submission. We weren't allowed to speak to other students, and the "teachers" focused more on discipline than on teaching.

One day, I became so bored I just wanted someone—*anyone*—to acknowledge my presence, so I raised my hand to go to the bathroom. It would at least give me something to do other than stare at my pencil. But no one called on me. So, I raised my hand a little higher, waved it around, and lifted it up with my other arm—trying desperately to get someone's attention. Still, no one acknowledged me.

On the front of my desk was a three-by-five notecard containing all my information—name, school, and grade. Eventually, I grew so fed up and impatient that I pulled a sheet of paper from my workbook and wrote, "SAVE WATER. DRINK BEER."

I peeled the tape off the notecard and swapped it with the sheet from my notebook. Then, I lowered my hand and waited for someone to notice.

When someone finally saw my handiwork, a voice came over the loudspeaker: "John Mathews, please report to the principal's office."

I stood up from my desk, but instead of going straight to the principal's office as instructed, I walked right to the bathroom. I could see the principal watching me as I passed his office.

I took my time in the bathroom, and when I finally made my way to the principal's office, his expression said everything I needed to know. "Mathews," he said, "Would you like to be suspended or paddled?"

I stared right back at him and replied, "I don't give a fuck what you do."

With that, he didn't utter another word. He walked over to his cluttered desk, casually sliding aside a stack of papers before placing his hand on a long, worn wooden paddle. As he picked it up, I could see its surface was marked by age and use, bearing the signs of countless stories—none of which ended well for those on the receiving end.

"Place your hands on the desk and lean forward," he ordered.

I did as I was told. He didn't give me just one beating—he gave me two, swinging with such force that my feet left the ground. The first lick sent a searing pain throughout my entire body, and my vision blurred as tears welled in my eyes.

By the second, the sting burned so deep I thought my skin would split open. But no matter how much it hurt, I wasn't going to let this man see me cry. I had taken worse beatings from my old man, and if I had learned anything from those, it was how to take a hit without letting it show. No matter how many times he struck me, he wasn't going to break me.

I walked back to the gymnasium, but the pain was unbearable. Every step sent a fresh wave of fire through my body, and when I reached my desk, I knew sitting down wasn't an option. The moment I tried, tears pricked the corners of my eyes. Instead, I stood behind my chair, shifting my weight, trying to find some relief.

But there was no sympathy here. A teacher glanced up, barely acknowledging my discomfort. "If you can't sit down," she said flatly, "then you can stand in the box."

The box.

It sat in the corner of the gym—a small, black plywood structure, just wide and tall enough for one person. Once inside, you weren't allowed to move. You weren't allowed to turn. You simply stood there, trapped in silence, waiting for time to pass.

I looked at my desk and then at the box. I had two choices: pain or darkness. I chose darkness. Being inside the box felt suffocating, and with every passing minute, my mind raced between counting down the seconds and questioning whether this was even legal—telling myself it was temporary.

But then, just like that box in the gym, the walls of ALC closed in tighter too. Initially, I was only "sentenced" to three days at ALC. So, on my final day, when I submitted my work packet, I said to the teacher, "Last day here."

She glanced at my name written on the folder and looked up. "Last day? Mathews, it says here that you're with us for another thirty days."

As I boarded the bus returning to high school, I was fuming. As soon as we arrived, I headed straight for the principal's office. When I walked in, the administrator said, "John, you can't be here. I can have you arrested for trespassing." Legally, I was not allowed to be back on school property. The principal heard the commotion from his office and stepped out to find out what the problem was.

"Oh, this won't take long," I said to both, "I'm dropping out." With only two months left until graduation, I felt vindicated as I turned to leave. Just then, I heard the principal's voice behind me say, "Good try, but we sent you to truancy court, so now you're ours until you turn nineteen."

Apparently, the judge had issued some creative sentencing aimed at keeping students in school until they turned nineteen. So, even though I was legally an adult, I couldn't drop out. The principal was right; I was essentially their property for another year.

The confrontation felt like a strategic chess match as I entered the office, intending to confront them for what I perceived as a cowardly decision to keep me in ALC for thirty days. It was a battle of egos—me, eighteen years old, facing off against a forty-five-year-old man, both of us determined to assert our positions for different reasons.

In hindsight, I recognized that dropping out of school would only provide a temporary victory for my ego, with long-term consequences. It was driven by my arrogance and the belief that I had power in the situation. On the other hand, *letting* me drop out would have had no impact on the man's life or career, yet he seemed just as determined to win as I was.

That arrogance and unwillingness to back down were traits I had inherited. It was the part of my father in me I didn't want to recognize. Dropping out might have felt like taking control, but in reality, it would have meant handing them the victory. So, the next day, I returned to ALC to serve out the remainder of my time—not because they had defeated me, but because I refused to allow my pride to rob me of what little future I had left.

This time, I put my head down and tried my best to make the most of it, or really, to make it out with no additional time or beatings. If I was stuck there, I might as well try. I even made a few friends and managed to avoid the wrong end of the paddle, and after thirty days, I returned to my regular school for the last week of my senior year. By this time, all my other classmates had left. Exams had been completed, and grades had been submitted for the year. Essentially, school was over.

I managed to scrape by with passing grades in all my classes except for one: English with Mr. Lee. Without a passing grade, I couldn't graduate high school. So, I made a bold decision to knock on Mr. Lee's door, hoping for a chance to make things right. As he waved me in from the other side of the glass, I nervously entered and approached his desk. I mustered the courage to ask him what I needed to do to achieve a "D" in his class so I could graduate high school.

Mr. Lee responded clearly and concisely, "You'll have to come in every day this week and catch up on all your unfinished work."

Determined to make things right, I committed to this plan. Each day that week, I arrived early and took my seat before class began. The room was cold and eerily quiet, with only Mr. Lee and me in it. But I stayed focused, diligently completing every assignment he handed me, one task at a time.

On the last day, Mr. Lee told me that my final assignment was to submit a five-page, typed book report along with an entire semester's worth of spelling words, their definitions, and a sentence using each word. Essentially, all the assignments I had opted out of completing throughout the year.

He informed me that only then would I pass the class and be able to graduate. That night, I went home and skimmed through a book just enough to write a paper. I also finished the assignment on spelling words. My wrist hurt from staying up all night to get the writing done, but I managed to complete the assignments.

The next day, I walked into Mr. Lee's classroom and handed him my assignment. However, instead of accepting it, he glanced down at the stack of papers and said, "The book report was supposed to be typed."

I explained that I didn't have a typewriter, a computer, or even a place to live; I was still sleeping on my friend's couch. He never said a word; he just stared me straight in the eyes and shrugged his shoulders.

I walked out of his room, feeling defeated. For the first time, I couldn't solve a problem, even if it was of my own making. Then, all of a sudden, a thought came to me. I ran down the hall to the only teacher in the building who might be willing to help me, Mrs. Luttrell. She always treated both Mike and me with love and respect. She even entrusted us with running the school store she operated out of her classroom, selling school supplies, breakfast sandwiches, and snacks to other students. When I told her what had just happened with Mr. Lee, she grabbed the book report out of my hand, sat at her computer, and typed out every word for me. She even corrected a few grammatical errors, ensured it was double-spaced, and printed it out before the final bell rang.

I sprinted back down the hallway to Mr. Lee's classroom with my freshly printed five-page book report in hand. Out of breath, I walked into his room and handed him my report. However, he wouldn't even take it from my hand.

With my arm still extended, he looked at the report and then back at me and said, "I'm not taking this. The report was due at the beginning of class, not at the end."

As he spoke, the sound of the period bell interrupted him, signaling the end of class. At that moment, I realized I had failed his class.

Twenty years later, I ran into Mr. Lee. At that time, I was the vice president of a one-hundred-fifty-million-dollar company, and when I walked into our shop building, he was there visiting an old friend of his. It was the first time I had seen him since that last day of school.

I approached him and said, "I don't know if you remember me; my name is John Mathews. I failed your English class twenty years ago due to unfinished assignments, and I've always wanted to know why."

A grin spread across his face as he replied, "Oh, I remember you. How could I forget? When I received your packet of work back from ALC, not only did you refuse to complete the assignments I gave you, but on one of the papers, you wrote, 'SAVE WATER. DRINK BEER.'"

Here was the truth all these years later: I was never going to pass Mr. Lee's class, typed paper or not. The man didn't intend to impart a lesson about responsibility or the importance of doing things right the first time. He wasn't trying to teach me anything—he was just pissed off, and honestly, I don't blame him.

Until now, he believed those words were directed at him. Little did I know that such a simple act of defiance at the ALC would cost me so much. Whether or not he meant it to be a lesson, it became one of the defining moments of my life.

Yet, it wasn't just one event that led to this moment; it was a series of poor decisions, each one compounding the next. Every choice we make—no matter how insignificant it seems—

has the potential to shape our lives in ways we can't foresee. In this case, it all started with a seemingly insignificant decision to leave school without following the rules, which landed me in ALC. From there, things spiraled downward. Frustrated by being ignored, I retaliated by scribbling "SAVE WATER. DRINK BEER" on a seemingly harmless sheet of paper. Ultimately, it landed me a beating and, worse, unintentionally made my English teacher so angry he would have me come in and do classwork without any intention of letting me pass his class.

Looking back, I realize that every encounter has the power to shape not only a single moment but the entire course of our lives. The lessons I learned at the ALC—and from those who showed me kindness when I least deserved it—serve as lasting reminders of the ripples we create.

Those ripples—whether driven by pride or compassion—extend beyond us, altering not only our own path but the lives of others. The wisdom comes from understanding that life often presents the test first and the lesson later, sometimes years or even decades down the line. The power has always been in our choices: we can either spread generosity, like Mrs. Luttrell and the stranger on the side of the road, or let ego and stubbornness dictate our actions, seeking validation at any cost.

At that time, the school administration, Mr. Lee, and I were all locked in a silent battle, each of us determined to prove something. None of us recognized the ripple effect we were setting into motion, too focused on winning a fight that, in the grand scheme of things, really didn't matter. Once the ripples start, we lose control, and despite our desire to stop them, we cannot. Part of the lesson, if not the entire lesson, is having to sit back and watch the destructive pattern run its course while wanting nothing more than for the water to return to a calm state, which, in this case, took decades to achieve.

But just as negativity compounds, so does kindness. With its natural pay-it-forward energy, even the simplest act of generosity can echo far beyond what we can see, creating a chain reaction with no clear end. Whether you refer to it as Karma or, as Christianity teaches, "We reap what we sow," the outcome remains the same: the ripples of kindness possess the ability to transcend time. And when these ripples of kindness gain momentum, they can turn into waves that stretch further than we could ever imagine.

Every choice we make sends out ripples that extend beyond our immediate world. Whether born of compassion or ego, those ripples shape the environment around us, determining whether we build or destroy.

Now I realize that true wisdom isn't just recognizing the weight of our decisions—it's acknowledging the lasting impact they create. It's a responsibility that requires both humility and foresight. The legacy we leave behind isn't solely defined by the moments we cherish but by the choices we make along the way. And if we choose wisely, if we choose to create ripples of positivity, we can help shape a better world—not only for ourselves but for generations to come.

CHAPTER 13

THE PERFECT STORM

Despite failing Mr. Lee's class, I refused to let that be the end of my story. Motivated and determined not to carry with me the label of a high school dropout, I enrolled in summer school to make up for the missing English credit. I was the only senior in the class, but that didn't bother me. All that mattered was earning my diploma, and when I finally did, I stood proudly as an official high school graduate.

After finishing school, I found a small apartment for Mike and me—a place we could call our own. I maxed out the last of my credit card to cover the deposit and buy some essentials to make it livable. Luckily, Mike was still on good terms with Dad and had been living at home, so he managed to bring over a hand-me-down couch and, more importantly, our beds and dressers. As I stood at the door, keys in hand, and looked around at the modest space we had created, I finally felt a sense of freedom and ease, as though life was offering us a fresh start.

Since Mike and I were the first of our friends to get an apartment, our place quickly became the go-to hangout spot. It wasn't always about doing drugs—most nights, we stayed up until dawn playing video games simply because we could, with no one around to tell us to turn it off and go to bed. Our couch became a revolving bed for friends who would come and crash while we went to work. The kitchen wasn't stocked with much beyond ramen noodles and peanut butter and jelly sandwiches, but to us, it didn't matter. We felt like real adults, convinced we had our lives together, even if the reality was far from it.

I was working as a full-time server in Gatlinburg. Since most of the customers coming into the restaurant were on vacation, I was making a lot of money, and it was all cash. This job allowed me to cover my rent and support my increasing drug habit. Every dollar I earned went towards bills or drugs, and when I didn't have enough to cover both, drugs always took priority. Paying bills would have to wait, and saving money was never an option.

What started as a curious teenager experimenting with new drugs eventually progressed. Over time, drugs became a regular part of my life, as well as Mike's, maintaining the parallel track we had both been on *together* since birth. Now, with the freedom of living on our own and a steady flow of cash, the door was wide open for more parties, more late nights, and, inevitably, more drugs.

On top of that, Tennessee was on the brink of an epidemic. The opioid crisis was looming as OxyContin, a powerful painkiller, made its way onto the market and was prescribed at alarming rates throughout the late 1990s and early 2000s. Pharmaceutical companies promoted it as a safe, effective solution for managing pain, while doctors, reassured by these claims, were encouraged to prescribe it freely. But Tennessee, like much of the country, was particularly susceptible to the drug's highly addictive nature—something that would soon reveal itself with devastating consequences.

Tennessee had high rates of opioid prescribing, with physicians writing prescriptions for opioids at a much higher rate than in previous decades. Between 1999 and 2015, the number of opioid prescriptions in Tennessee increased by 415%.[iv] This surge in prescriptions set the stage for a devastating public health crisis that would grip the state for years to come.

Exacerbating this growing problem, Tennessee was also home to several so-called "pill mills." Clinics that inappropriately provide opioid prescriptions. These pill mills often employed doctors who were willing to write prescriptions for opioids without proper medical justification, which contributed to the proliferation of opioids in the state.

But perhaps what made Tennessee particularly vulnerable was its underlying socioeconomic landscape. Low income and limited educational opportunities are among the primary risk factors for substance abuse[v], and Tennessee had both. Additional factors like job loss and limited access to affordable healthcare further heightened the risk, creating a perfect storm. For me, these societal pressures, combined with a family history of substance abuse and early exposure to drugs, would eventually converge, derailing my life in ways I couldn't yet imagine.

But in 2000, OxyContin wasn't on my radar. I had no idea that I was standing in the crosshairs of one of the worst epidemics the country would ever face. Tragically, my experiences, trauma, and escalating drug use made me a prime candidate—a perfect poster child for what substance abuse could do to a life.

The warning signs were there—the freedom of living on my own, the flow of easy money, and a state drowning in a wave of opioids waiting to crash. But when you're living it, you don't always notice how quickly the water is rising. I had convinced myself that I had things under control, but the reality was that addiction was already circling me, tightening its grip without me even realizing it. And then, one small yellow pill would come into my life and change everything.

CHAPTER 14

THE ADDICT BRAIN

Each time I tried a new drug, I would compare it to the others, with each experience being different. For the most part, I just really enjoyed being high. There were a few drugs I didn't care for, like Xanax or any downer, for that matter. I preferred the ones that gave me energy, the ones that made me feel alive. The only exception to this rule was marijuana. I smoked weed every day and actually preferred it over alcohol, but I wasn't your typical pothead.

I had become a functioning drug user the way many people become functioning alcoholics—I could still show up to work, go through the motions of daily life, and give the appearance of having things under control. I had suffered through acute depression and anxiety for as long as I could remember, but the drugs dulled the edges of those emotions. They numbed me, shielding me from having to feel any complex emotions - providing me with an escape from myself.

Eventually, you stop asking yourself why you need the drugs. It just becomes another part of your routine, like brushing your teeth or setting an alarm. At some point, you don't even think about it—it's automatic, muscle memory. And with no immediate consequences, there's no reason to sound an alarm or question the habit. But then, OxyContin entered my life, and in an instant, everything changed.

I was at a friend's house when he pulled out a bottle of pills I had never seen before. They were small, round, and coated in a dull yellow exterior. The letters "OC" were stamped on one side and the number "40" on the other. He explained that they were pain pills—like Percocet, but much stronger—and that crushing and snorting them would deliver a faster, more intense high. This was my first encounter with OxyContin, the drug that would soon take over my life.

At that time, OxyContin was available in 10mg, 20mg, 40mg, and 80mg tablets, and by the year 2000, Purdue Pharma, the company that marketed and manufactured the drug, introduced a 160mg pill.[vi]

To help identify the doses, each OxyContin had a distinct color: the 10mg was white, the 20mg was pink, the 40mg was yellow, the 80mg was green, and the 160mg was blue. In comparison, Percocet, the commonly prescribed painkiller for minor injuries, contains 5mg of oxycodone. Therefore, to match the amount of Oxy we just snorted up our nose in seconds, you would have to take eight Percocet. I didn't know it then, but those small, color-coded pills would become the center of my world—and my eventual undoing.

OxyContin was more expensive, more potent, and far more addictive than anything I had ever encountered. Its high was unlike any other drug I had ever tried. Almost immediately after the drug entered my system, my mind slipped into an intense and euphoric state, giving me a feeling of extreme happiness and relaxation. It was a rush of pleasure I had never experienced before, and it had a cascading effect on my mind.

I was Superman on this drug: I felt smarter, worked harder, and had a seemingly endless supply of energy. I felt like I could do anything. In 2019, *Time Magazine* wrote an article called "The Science of Addiction," which states, "When exposed to drugs, our memory systems, reward circuits, decision-making skills, and conditioning kick in salience in overdrive - to create an all-consuming pattern of uncontrollable craving." This particularly applies to OxyContin, "The brain loves OxyContin - the way the drug lights up the limbic system, with cascading effects through the ventral striatum, midbrain, amygdala, orbitofrontal cortex, and prefrontal cortex, leaving pure pleasure in its wake. What the brain loves, it learns to crave."[vii] What began as a superpower would later become a prison as the drug's grip tightened and the illusion of control crumbled.

When Purdue Pharma was first criminalized in 2007 for misleading doctors, regulators, and patients about the drug's addiction risk, the nation's attention was finally drawn to the growing opioid crisis. What was once dismissed as an isolated problem had become a full-blown epidemic. Today, there is an abundance of knowledge surrounding the devastating impact of OxyContin, but that awareness came too late for millions of lives that have already been, and continue to be, destroyed. My story is just one in a vast ocean of suffering, a personal narrative woven into the collective tragedy of a nation gripped by addiction.

The day after I first tried the drug, my first thought the following morning was, *"How can I get more?"* Oxy was still new to our area, and most of my friends hadn't heard about it yet. I no longer craved the hydros I used to take, which was unfortunate because I could get them anywhere. Eventually, I was able to find a steady supply of Oxy, and since it didn't take much to get high, a couple of pills lasted me a few days. But that wouldn't be the case for long.

As time went on, it took more and more Oxy to achieve the high I craved, which meant more pills and more money, which was disappearing faster than I could keep up. What had started as a pursuit of euphoria quickly spiraled into an insatiable need—a craving that emptied my pockets and consumed my thoughts. I had always believed the word "addict" was meant for junkies, meth heads, and heroin users—the people living under the bridge, not someone like me. Even as my dependence on Oxy grew and my daily life revolved around the next pill, I couldn't see the truth staring me in the face: by definition, I was becoming addicted.

It crept up on me like a slow-moving train—steady, inevitable, and impossible to stop. I convinced myself I was fine, but that was the voice of what I now call the "Addict Brain." This part of the brain is cunning, persuasive, and relentless. Its sole mission is to override logic and convince you that you need more—whether it's a pill, a drink, or a sugary treat. The Addict Brain doesn't reason; it persuades. It's the voice that whispers, *"You've had a rough day; you deserve to have one; or one more won't hurt."* And the scariest part? It lives in all of us.

However, what the Addict Brain craves depends on its host. For some, it's sugar, whispering, *"You've been good; you deserve a treat."* For others, it's a drink, a gamble, or a pill. The Addict Brain doesn't care about long-term consequences—it thrives in the immediate, promising relief in exchange for self-destruction. What makes it so dangerous is its ability to disguise itself as reason, convincing you that giving in is a reward, not a step toward dependency.

In most cases, the Addict Brain wins. Maybe not immediately, but the Addict Brain is patient—it doesn't need to rush. It works quietly in the background, lurking in the shadows, waiting for the right moment. It knows every one of your weaknesses, every crack in your armor, and it will exploit them relentlessly. No matter how long it takes—hours, days, weeks, or even years—it will wear you down until you finally give in and surrender to the craving you thought you could resist.

At the time, my Addict Brain had me fully convinced I was a better bartender when I was high on OxyContin. I told myself it enhanced my personality, made me more outgoing, and turned me into the kind of person who could rake in bigger tips. So, I made sure to get high before every shift. After all, I *deserved* to make as much money as possible. The Addict Brain whispered that everything was fine: I had a job, a social life, and I wasn't a junkie under a bridge. I wasn't an addict. But denial doesn't change reality. Just because I wouldn't admit it didn't mean it wasn't true. I was, without a doubt, an addict.

As a bartender, I made good money—enough that, if I hadn't been using, I could have built some stability. I could have started a savings account, fixed up a reliable car, or gotten a place of my own. But that wasn't my reality. After a good night behind the bar, my mind didn't think about future plans or financial security. It went straight to one thought: *How many pills can this money buy me?* At the height of my addiction, the goal wasn't even to get high anymore. The pills had become my lifeline, the thing I needed just to make it through the day.

I could function with pills, but without them, I was a mess. My body had developed more than just a chemical dependency on OxyContin—it was a physical one that demanded constant feeding. There were moments when I would catch a glimpse of myself in the mirror, and the reflection staring back was almost unrecognizable. The man in the glass looked tired and worn, with hollow eyes and a face drained of life. His skin seemed stretched too thin, his once bright expression dulled by exhaustion and addiction. I would stand there, frozen, trying to reconcile who I saw with who I believed I still was. For a brief second, my logical brain would resurface, sending a jolt of fear through me. It was just enough to make me think, *I need to quit*—but the Addict Brain always had other plans.

That resolve was always short-lived. By the following day, I would wake up in pain—the first signs of withdrawal creeping in, dragging me down before I could even get out of bed. My body was in chaos, with every autonomic system misfiring as it desperately tried to return to homeostasis. Even the smallest movement felt like agony. And just then, right on cue, the Addict Brain would start its shit: *"Just take another pill, John, and all this will go away."* In that moment of desperation, I'd surrender. I convinced myself, once again, that as long as I had enough money and the pills kept coming, I didn't have a problem. I could take a pill every day and avoid this misery. It seemed like a solution and not the trap it really was.

The Addict Brain made it easy to avoid confronting the truth of my addiction. It provided excuses, distractions, and justifications, shielding me from the reality that my dependence on OxyContin had gone beyond concerning—it had become dangerous. By this point, I was taking six to eight 80mg pills a day, the ones we called "Green Beans" because of their distinct green color. To put that into perspective, that's the equivalent of swallowing a hundred Percocet pills—the same ones your doctor prescribed after that minor surgery—in one day: every day. I knew it was insane, but the Addict Brain had a way of making it seem normal.

At this stage of my addiction, OxyContin wasn't just something I craved—it was the only thing that gave me any sense of satisfaction. The more I used it, the more my brain began to reject the natural joys that once made life worth living. Eating a good meal, hanging out with friends, or listening to music that used to move me now felt dull and meaningless. The Addict Brain had narrowed my world to the point where my only source of reward came from those green pills. It's something scientists call the narrowing of the reward system[viii]—when your brain becomes so dependent on the surge of dopamine from drugs that nothing else brings you joy. The rush from Oxy had hijacked my reward system completely. My brain was constantly screaming for one thing and one thing only: more Oxy. Without the pills, life felt empty. And with them, life was controlled chaos.

But the scariest part was that I didn't realize this was happening. I just thought life was becoming less exciting, less rewarding. What I didn't understand was that my brain had recalibrated itself to believe that Oxy was the only thing capable of giving me happiness. In reality, it was dragging me deeper into a pit where nothing else mattered.

To add to the problem, I discovered a new, more powerful way to ingest the drug—a needle. Snorting the pills had once been enough to numb my body and mind, but that no longer satisfied the cravings that consumed me. I had crossed another line, one I never thought I'd approach, and with that needle, the Addict Brain tightened its grip even further. The spiral was accelerating, and the deeper I fell, the harder it became to see a way out. What started as an escape had now become my prison.

CHAPTER 15

WHAT COULD GO WRONG

The first time I used a needle, it felt almost as if fate had orchestrated the moment—like every misstep and poor decision quietly steered me here, to this point of no return. I wasn't actively searching for a new way to get high, but the Addict Brain doesn't need you to search. It's patient and opportunistic. When the right moment presents itself, it knows exactly how to make you say yes, whispering reassurances that feel like destiny. That night, at a friend's house, we were partying as usual—drinking, smoking, and snorting Oxy. I had just taken a line off the coffee table when a buddy pulled me aside, his expression halfway between excitement and secrecy.

"Come on, I want to show you something," he said, nodding toward the kitchen.

I followed him, curiosity outweighing any instinct I had to turn back. There, on the granite countertop, sat a setup that looked deceptively simple: a spoon, a few cotton balls, and a syringe.

Ordinary objects, yet they might as well have been weapons. The spoon gleamed under the fluorescent light like a sharpened blade. The syringe, its plastic barrel barely noticeable at first glance, looked sterile and lifeless, waiting for its chance to leave a mark.

I watched as my friend methodically crushed an Oxy on the countertop, just as we'd done countless times before. But this time, he wasn't preparing a line to snort. He carefully funneled the powder into the spoon and added just enough water to cover the surface. Holding the spoon steady in his left hand, he pulled a lighter from his pocket with his right, placing the flame directly beneath the spoon. The powder slowly dissolved into the water, leaving behind a thick, warm liquid.

He tore a small piece from one of the cotton balls, pressing it into the liquid. Then, with a precision that came from practice, he guided the needle tip through the cotton, filtering the solution as he drew it into the syringe. He tapped the barrel with his middle finger, releasing any air bubbles. I could hear my own heartbeat pounding in my ears, part of me screaming to leave, but I stood rooted to the floor—part horrified, part entranced. I was witnessing a ritual; one I knew deep down would soon become my own.

He rolled up his sleeve, tracing the veins in his arm before sliding the needle into place. As he pushed down on the plunger, I saw the transformation hit him like a wave. His pupils shrank into tiny pinpoints, his limbs went limp, and a lazy, euphoric smile spread across his face as he slumped back onto the nearby couch. The person he had been just moments earlier was gone, replaced by a version of himself that looked at peace as if nothing in the world could touch him.

"Hey Mathews, you wanna try?" he asked casually as if he hadn't just crossed a line that could never be uncrossed.

I stood there, staring at the syringe still resting in his hand, my thoughts racing. I had promised myself I'd never go this far. *I wasn't a junkie—I was better than that, right?* But the Addict Brain has a way of dismantling your moral compass, piece by piece, until the things you swore you'd never do start to feel inevitable, even logical. All it took was one whisper: *We've always wondered what that feels like. One time won't kill you.*

That "once" didn't stay once. The needle quickly became my new normal, and with it came new lies, new layers of shame, and a desperate need to hide how bad things had gotten. I started hiding how severe my OxyContin addiction had become. Few friends knew the extent of it, and even fewer knew that I had started injecting the drug.

At first, I kept it from Mike, who was struggling with his own substance abuse. I should have protected him from it—I should've been the person who warned him, not the one who led him down a darker path. But the truth is, I didn't want to feel alone. The Addict Brain isn't just selfish with drugs; it's selfish with the company it keeps too. So, I convinced Mike to try it with me. Looking back, I can see how shitty that was. Introducing something so awful to him wasn't just careless—it was cruel. But at the time, I must have subconsciously wanted him to like me more, to join me, so that *I* could feel better about myself and my own choices. Luckily for Mike, his Addict Brain wasn't in complete control. That first time ended up being his last. He never picked up the needle again.

Mandy was a different story. I hid everything from her for as long as I could. She shared the same circle of friends and some of the same habits, but I knew this was a line I couldn't let her see me cross. I wore long-sleeved shirts when we were together so she wouldn't notice the bruises tracking their way up my arms, and I told her I'd started exercising to explain away my weight loss.

If she suspected the truth, she never let on. Either she believed my lies, or she was holding onto her own denial. As for me, I was simply surviving—getting myself from one day to the next, from high to high, and always from one supply to another.

I had driven away most of the people who had once cared about me. My mom, who had always been my biggest supporter, had finally reached her breaking point. After years of watching me lie, steal, and manipulate, she cut me off—financially and emotionally. I was on my own, and part of me knew I deserved it. But that didn't stop the downward spiral.

My habit was becoming increasingly harder to sustain. I was burning through money at an alarming rate, and with OxyContin's popularity surging, the suppliers saw an opportunity to profit and began hiking up their prices. Each pill cost more than the last, stretching my resources thinner by the day. And then, the worst thing possible happened. My dealer went to jail, and suddenly, I wasn't just dealing with a cash flow problem—I had a full-blown supply crisis. For someone like me, dependent on the pills just to feel normal, this wasn't an inconvenience. It was a threat to my survival.

I *needed* the pills. There was no other option. Money or not, supplier or no supplier, I was determined to find a way to get them. That's when I started stealing. It began small—lifting items from Walmart and pawning stolen equipment from people's yards. But theft, much like addiction, is a slippery slope. One bad decision led to another, and soon, I found myself doing things I never imagined.

I began forging checks from my deceased grandmother's bank account, writing them out to myself, and then cashing them. In the end, I had stolen over fourteen thousand dollars from her account. Just like my drug habit, stealing became easier over time.

My Addict Brain was sharpening its skills, fine-tuning its ability to justify everything I did. When I stole, I didn't feel like a criminal—I felt like someone doing what was necessary. I needed the drugs to function, and that single thought was enough to rationalize nearly anything. Every wrong move I made came with an internal stamp of approval: *You're just doing what you have to do.*

By the time I realized how much damage I had caused—to myself, to my family, to anyone who tried to help me—it was too late. The lies, the theft, and the betrayals had already left their mark. But the scariest part was that I didn't care. The only thing that mattered was finding the next high; nothing else registered beyond that. Deep down, though, I knew my luck would run out. And when it did, I wouldn't be able to control the fallout.

CHAPTER 16

FIRST OF MANY

The first time I got arrested, I was in a car with three of my friends, cruising through downtown Gatlinburg on a Friday night. The plan was simple—hang out, smoke a little weed, and enjoy the weekend.

When we turned onto the strip, traffic was at a standstill, packed with tourists and locals inching their way down the main Parkway. With nowhere to go, we rolled down the windows and finished the joint we had lit earlier, hoping the breeze would carry away the smell. Then, we lit cigarettes, thinking we could mask any lingering odor. We weren't exactly being discreet—laughing, blasting music, and joking like we were untouchable.

We didn't notice the bicycle cop weaving through traffic until he passed our car. For a moment, I thought we were safe. We weren't actively smoking anymore, so the smell wasn't as strong. But about one hundred feet ahead, he suddenly whipped his bike around and sped toward us. My heart pounded.

He pulled up beside us, got off his bike, and approached my window. "License and registration," he said, his tone making it clear he wasn't buying our cigarette coverup. I handed him what he asked for, trying to keep my face neutral. After a moment, he pointed toward the side of the road. "Step out of the vehicle, please."

We all climbed out, and I watched him search the car, moving methodically from the front to the back seat, his flashlight sweeping every surface. My buddy, standing next to me, shifted nervously. As the officer moved toward the back of the vehicle, my friend slid his hand into his pocket, pulled out the small bag of weed, and dropped it on the street. Then as casually as he could, he stepped on it and pressed it onto the pavement.

For a moment, relief washed over me. I thought we had gotten away with it. But when the officer turned back toward us, he noticed my buddy's foot awkwardly planted on the ground. "Lift your foot," the cop said, narrowing his eyes. My friend hesitated but eventually lifted his foot, revealing the bag crumpled beneath his shoe.

The officer smirked as he reached down, picked up the bag, and held it up like a trophy. "Who does this belong to?" None of us answered. We didn't need to. I was driving the car, and in the cop's eyes, that made it mine. "Turn around and put your hands behind your back," the officer said.

I didn't argue. I heard the metallic snap as the first cuff tightened around my wrist, then the second. The cold steel bit into my skin as the officer led me toward the squad car that had arrived for backup. I could feel the weight of everyone's eyes on me—the tourists, the locals, even the guy selling taffy at the corner store. It felt like the whole town had stopped to watch me get arrested.

In the back of the police car, I shifted uncomfortably, trying to find a position where the cuffs didn't dig into my wrists. The silence inside the car was deafening, broken only by the hum of the engine. I tried talking to the officer, hoping for a shred of acknowledgment—something to remind me that I was still human. But he didn't even glance in my direction.

When we pulled into the station, the door opened, and I was taken into the booking room. The process was mechanical and impersonal: fingerprints, mugshots, and a brief stop before they led me to a holding cell. Four or five other guys sat on the concrete bench along the back wall, none of whom seemed particularly interested in my arrival. A shiny chrome toilet without a seat gleamed under the fluorescent lights in the corner.

I sat down cautiously, trying to keep my trembling hands hidden from view. My mind raced with thoughts of what might happen next. My only experience of jail came from television shows and movies—images of cellmates fighting, hazing the new guy, or ganging up in groups. I felt like there was a flashing neon sign above my head that read, "Fresh Blood."

But as the hours passed, I realized those fears existed only in my head. No one so much as even looked my way. The only sound was the distant clanking of doors and the occasional cough from another inmate. I kept my head down and my eyes open, counting the minutes until morning.

Early the next day, I was led into a small room where I sat with my fellow inmates. The magistrate called each of us one by one to read off our charges and set bail. When my name was finally called, I shuffled into the room, still groggy from the night before. The magistrate barely looked up as he read the charge: "Simple possession. Bond is set at one thousand dollars."

Afterward, they took me to a payphone mounted on the wall next to a bulletin board covered with business cards from local bail bondsmen. I guess this was my "one phone call." I scanned the cards and picked one: Ace Bonding. From that day forward, they became my go-to bondsmen, and I called them every time I landed behind bars.

The woman who owned the company always treated me with respect and kindness—never a lecture, never a judgmental glance. In a world where most people had already written me off, that small, simple kindness was rare.

This experience soon became all too familiar. At first, the arrests were for minor infractions like driving on a suspended license, possession, or public intoxication. But soon, the charges escalated—like aggravated assault and criminal trespassing, which I earned after throwing a man over the counter at Chick-fil-A.

In the heat of the moment and high on drugs, I lunged after him, shoving workers out of my way to finish the fight. By the time the police arrived, I was long gone. But someone had written down my friend's license plate, and it didn't take long for the cops to track him down and question him. Once they had my name, a warrant was issued for my arrest. Eventually, I turned myself in and posted bail.

These incidents became part of a pattern, and before long, I was racking up arrests and spending nights in holding cells across East Tennessee. I quickly learned that no two county jails were alike. Some were large and well-organized, with strict routines and little room to breathe. Others were small and chaotic, making them easier to navigate and, in some cases, easier to handle. Jail wasn't just a punishment or an anomaly anymore—it had become part of my life.

One night, while in a small holding cell, I shared a bunk bed with a man who I thought was ignoring me at first. I assumed he was asleep—until he eventually opened his mouth to speak. But what came out wasn't speech. It was garbled, unintelligible mumbling that made no sense. It didn't take long for me to realize the man didn't have a tongue.

With no way to speak, he used his fingers to trace letters on the bed, spelling out words one letter at a time. Since we had nothing to physically write on, his messages would fade as fast as he created them, but it was enough to keep the conversation going. With nothing but time, we sat there, communicating for hours.

I never found out how he lost his tongue, but judging by the scars on his face and the haunted look in his eyes, I could only imagine it wasn't voluntary.

On a different night, I found myself locked up with an inmate about three times my size. His tattoos covered every inch of his body, and his eyes had a vacant, unsettling look that made my skin crawl. There was something off about him, something you couldn't put into words but could feel. Some would say he had "crazy eyes."

He paced back and forth in the cell, humming Pearl Jam's *Last Kiss* over and over, like a broken record stuck on a loop. His footsteps were methodical, and his focus seemed distant as if he were somewhere else entirely. At first, he hardly seemed to notice me, and I hoped it would stay that way. But after a few hours of pacing, he suddenly stopped and stared directly at me.

"Hey, do you know this song?" he asked. His tone was casual, but his eyes told a different story. I couldn't read him, and that made me uneasy. "Yeah," I replied cautiously. "That one by Pearl Jam, right?"

He nodded, and we ended up talking. He seemed almost normal at times—charismatic, even—but that underlying unease never went away. It was as if he were teetering on the edge, and I didn't want to be the reason he fell off. He reminded me of someone who might be suffering from schizophrenia, possibly off his medication for days. His mood would shift without warning, and every word I spoke felt like stepping onto thin ice.

Later, I found out he had been arrested that night for stabbing someone to death. The revelation sent a chill down my spine, and I couldn't stop thinking about the lyrics he had been humming all night:

"Oh, where oh where can my baby beee..." and *"The Lord took her away from meee."*

I couldn't help but wonder—was he singing about the woman he had killed? Was she the "baby" in the song? Either way, he wasn't exactly the kind of person you wanted to be locked in a cell with.

After spending enough time behind bars, I owed more money to bail bondsmen than I could afford. Thankfully, the owner of Ace Bonding liked me, and she allowed me to pay off my debts in installments. She kept track in a ledger, noting my balance in the red column like a bookie keeping tabs on a gambler. In the end, I paid her back every penny, though not without racking up more arrests along the way.

With all of the arrests came probation, which should have made things more difficult, but I found ways to work the system. Each time I was arrested, it should have triggered a probation violation, but somehow, I kept slipping through the cracks. At one point, I was on probation in two different counties, and neither of them knew about the other.

Each month, I had to check in with both probation officers, pass a drug test, pay a fee, and prove I was employed. Since I was still using, I relied on a friend's urine to pass the tests. Every visit was like playing Russian roulette. I prayed they wouldn't discover the lies or the double probation, knowing that if they did, it would be a one-way ticket back to jail.

At twenty-one years old, I had a criminal rap sheet, a pile of debt, and a life that was spiraling completely out of control. But none of that mattered to me. My only concern was the same as it had always been: how and when I could get my next fix.

These incidents didn't feel like isolated mistakes anymore—they were part of a pattern; one I couldn't seem to break. My time in holding cells, courtrooms, and probation offices blurred together like a bad dream I couldn't wake up from. I wasn't just losing control anymore—I had already lost it.

CHAPTER 17
ROCK BOTTOM

To this day, I've never experienced an urge as overwhelming as addiction. It's as if two different people were living within me: the person I was at my core and the monster that took hold—an uncontrollable, insatiable creature that thrived only on my destruction. Once that monster was unleashed, nothing was sacred anymore. It devoured everything: my dreams, my relationships, my future. Everyone in my life became a potential victim of its hunger. Nothing was off-limits.

The monster wasn't just a fleeting thought or a momentary craving; it was the Addict Brain fully evolved, an unrelenting force in survival mode. Addiction rewires the human brain to believe that the drug we crave is not just something we want; it is the very thing we *need* simply to survive.

Imagine the pang of hunger, but not the kind you get from skipping a meal or even missing a day's worth of food. This is the hunger you feel after going an entire week without eating—a hunger so primal that your body starts to shut down, your mind becomes foggy, and even the thought of eating spoiled scraps from a garbage can seems reasonable. At that point, dignity disappears, and survival instincts take over. Now, multiply that level of desperation a hundredfold, and you start to understand the addict's craving.

Without the drug, your body feels as if it will collapse, and your mind like it will fracture. You won't survive the next hour, let alone the next day. It doesn't matter if you're starving, sleep-deprived, or on the edge of ruin—none of those register. Just as starvation drives you to food to stay alive, the Addict Brain compels you toward the drug with the same singular purpose: survival at any cost.

But the addict's hunger is far more deceptive. Unlike a starving person whose hunger can be relieved with food, this craving can never be fully satisfied. One dose provides temporary relief, but as the high fades, the monster returns—stronger, louder, and more desperate than before. Addiction rewires the brain, altering its reward system so thoroughly that the drug becomes the only source of relief. What once might have been a choice—taking that first hit—disappears entirely. It transforms into an instinct, a reflex, driven by altered brain chemistry that convinces you that the drug is the only thing keeping you alive.

This understanding is crucial, especially for those who have never faced addiction. Friends and family members often find it difficult to grasp why someone they love—someone who is clearly suffering—continues to use. From their viewpoint, quitting appears to be the obvious answer. However, the logic of a rational mind loses its power when the Addict Brain takes charge. Instead of stopping, addicts—people like me—cling to the very thing destroying them because their brains have been rewired to perceive survival through the lens of addiction.

According to the U.S. Food and Drug Administration, opioid addiction has a relapse rate of 65-70% and is responsible for 75% of all overdose deaths in the U.S. Research also indicates that individuals with long-term addiction or multiple relapses face an even greater risk of relapse.[ix] I understood those statistics weren't merely numbers—they were my reality. I had attempted to quit countless times, but no amount of promises or willpower could conquer the monster that had taken control of me.

I knew I needed help. My friend Matt, with whom I regularly shot Oxy alongside, had come to the same grim realization. We were in bad shape, and unless something changed, we were both on a one-way trip to prison or death.

Matt took the first step by researching treatment options and discovered a medication called methadone. It could only be prescribed by a board-certified doctor and was part of a treatment method known as Medication-Assisted Treatment (MAT). At that time, in the early 2000s, methadone was the only widely available MAT option.[x] It had primarily been used to treat heroin addiction, but the clinic in Knoxville had started prescribing it to patients addicted to OxyContin as well.

Methadone works by binding to the same opioid receptors in the brain that drugs like heroin and Oxy target. What distinguishes methadone is its long half-life and high affinity for these receptors, meaning it binds strongly and remains in the system longer.[xi] This provides relief from withdrawal symptoms without creating the euphoric high associated with addictive opioids. Methadone essentially occupies the receptors like a gatekeeper, preventing other opioids from attaching and triggering a high. By stabilizing brain chemistry and easing cravings, methadone helps patients regain control over their lives.

Today, there are more MAT options available, including buprenorphine (with brand names like Suboxone and Subutex) and naltrexone (brand name Vivitrol). Suboxone acts as a partial opioid agonist, meaning it activates opioid receptors but to a lesser degree than methadone, which reduces cravings without causing the same intense effects. It has a ceiling effect that makes overdose less likely.

In contrast, Vivitrol serves as an opioid antagonist. Rather than activating the receptors, it completely blocks them, preventing any opioid from producing a high.[xii] However, Vivitrol is generally used only after detox, as it does not manage withdrawal symptoms like methadone and Suboxone do.

Matt tried to convince me to go to the clinic with him, but I was stubborn. I didn't want to "replace one drug with another," so I brushed him off, convinced I could kick the habit on my own. While Matt checked into the clinic, I continued using.

Unfortunately, I can't remember how many times I relapsed or how often I swore, "*This is it—I'm quitting for good this time.*" My periods of sobriety were short-lived, sometimes lasting just a few days, occasionally weeks, and on rare occasions, even a month. But what I do remember with painful clarity is what always pulled me back: my old friend, the Addict Brain.

It was slick, persuasive, and always waiting for the right moment. "*See? You **did** quit. You **are** in control,*" it would whisper. "*Quitting wasn't **that** hard. You can handle just one more.*" And every time I gave in, it strengthened its grip on me, pulling me deeper into the cycle I swore I'd escape.

Several months later, Matt came to our apartment. He looked cleaner and healthier than I had seen him in years. He seemed happy, full of energy, and had a brightness about him that I had never noticed before. The color had returned to his face, and his eyes had life behind them—eyes that no longer carried the weight of addiction. He didn't need to ask if I was still using because one glance at me told him everything he needed to know. I was a living reminder of the life he had just escaped. Instead, he asked me, "Do you have one for tomorrow?"

Every junkie knows exactly what that means. No matter how high you might be at that moment, your mind is always racing ahead, planning, scheming, and calculating how to get your next fix. His question wasn't about tonight; it was about survival—tomorrow's survival. And at that moment, I knew I didn't have an Oxy for the morning. I had already spent the whole day scrambling to figure out how to get one, a familiar loop that had become the soundtrack of my life.

But before I could respond, Matt said, "John, I made an appointment for you at the methadone clinic on the first day I went, and tomorrow morning is your assessment day."

At that moment, neither of us knew it, but those words would ultimately save my life.

Getting into the clinic was no easy task. You couldn't simply walk in off the street and expect to receive treatment. You needed to get on a waiting list that, at times, extended for months—just one of the many hurdles that addicts encounter when seeking help. The waiting could be a death sentence for someone trapped in the cycle of withdrawal and relapse. I didn't realize how fortunate I was that Matt had cared about me enough to think ahead.

On assessment day, you meet with the doctor to determine your therapeutic dose. Methadone isn't a one-size-fits-all solution; the goal is to find a dose that prevents withdrawal symptoms without inducing a high. The average dose ranges from sixty to one hundred twenty milligrams. Based on my drug use and history of dependency, the doctor prescribed me ninety milligrams—a dose intended to stabilize me, though at that moment, I wasn't sure if stability was even possible.

Matt offered to drive us to the methadone clinic in the morning. However, there was another issue—I didn't have medical insurance, so I would need to pay out of pocket. The treatment would cost one hundred thirty dollars a week. I didn't even have enough money for a single OxyContin, let alone enough for treatment. Panic set in as I realized my chance to get clean might slip through my fingers before it even began.

I didn't know what else to do, so I called my mom—the one woman whose trust I had betrayed time and time again, whom I had disappointed, robbed, and hurt. My heart pounded as I took out my phone, fully aware of the pain I had already caused her. I took a deep breath and dialed her number. The phone rang once, then twice. My heartbeat quickened with each unanswered ring, and for a moment, I thought she wouldn't pick up. But just as I was about to give up, I heard her voice on the other end of the line.

"Hey, Johnny Cakes," she said, in a warm and familiar tone. My grandmother had given me this nickname, which only Mom and Grandma ever used. Hearing it felt like stepping into a memory from a simpler time, before addiction had taken everything away.

My mom had every reason not to answer my call and even more reasons not to trust me. But there I was, clinging to the hope that she would listen. I worked up the courage to say what I had been practicing for the last hour before I called.

"Mom," I began, my voice cracking under the weight of guilt and desperation. "I know you have every reason not to believe me, and I know I've tried more times than either of us can count to get off this drug. But I have an opportunity to get clean—really get clean this time."

The phone went silent, and the pause felt like it stretched on forever. I held my breath, preparing myself for rejection. I wouldn't have blamed her if she had said no. She had every right to. But then, from the other end of the line, I heard her say the words that would change my life: "What do you need?"

Call it a mother's intuition or divine intervention, but something convinced her that I was telling the truth. Despite everything I had done to hurt her, disappoint her, and betray her trust, she gave me the money. She had every justifiable reason to hang up that phone and walk away, but she didn't. That day, her faith in me—when I had lost all faith in myself—saved my life.

At 5 a.m. the next morning, Matt picked me up, and we drove to the methadone clinic in Knoxville. When we arrived, we entered a room full of other addicts like us. It was a strange, sobering blend of humanity: the homeless man who probably slept under the bridge last night and the banker adjusting his tie on the way to work. Addiction didn't care who you were—it leveled us all the same.

The clinic felt like a depressing, high-security bank. A security guard stood in the corner, arms crossed, scanning the room as if he were anticipating trouble, as if one of us might rob the place. The lobby served as the waiting room. It felt sterile, was brightly lit, and was filled with people avoiding eye contact.

I received a number to protect my identity and was told to wait until it was called.

I sank into one of the hard plastic chairs next to a woman holding a baby. The child was bundled in a stained onesie, its tiny hands clutching the collar of its mother's shirt. You could tell that neither of them had seen a shower in a long time. The woman stared down at her cracked, dirt-covered feet, barely protected by broken sandals, her toes poking through the torn rubber. She never looked up. When her number was called, she slowly stood, shifting the baby to her other hip, her eyes still glued to the floor as she shuffled toward the entrance to receive her dose of methadone.

I turned my eyes to the ground, unsure whether it was from my own shame or out of respect for the others in the room. But in that moment, I realized we were all hiding something. The shame we carried didn't need words; it filled the room like a thick, unspoken fog.

After a while, my number was finally called, and I was ushered through a wooden door. I walked in and quietly closed it behind me, the soft click of the latch feeling heavier than it should. A glass barrier separated me from the nurse on the other side—an invisible wall of mistrust built by years of addicts trying to game the system.

"What's your name and dose?" she asked, her eyes glued to a sheet of yellow notebook paper as she scribbled notes.
"John Mathews, ninety milligrams," I said, my words sounding

more robotic than I intended. She didn't react, just nodded as if she had heard my story before.

Then, without saying a word, she handed me a cup filled with juice, the thick, chalky wafer of methadone dissolving inside. The wafer reminded me of the communion bread we used to take on Sundays—except this wasn't about forgiveness or salvation. This was about survival. The mixture was surprisingly sweet as it slid down my throat, coating my mouth with a gritty residue.

"Open," the nurse said, watching closely as I opened my mouth and lifted my tongue. They didn't take chances here. The process ensured I wasn't hiding the methadone in my mouth to spit out and sell on the street later. Trust didn't exist in a place like this.

I swallowed hard, the reality settling in. This wasn't just a dose of medication—it was a reminder that I couldn't lie or cheat my way out of this fight anymore.

I repeated this process day in and day out for about a year. The methadone didn't fix everything, but it allowed me to stop obsessing over the daily search for an Oxy and slowly start piecing my life back together. It wasn't an overnight transformation—it was slow, grueling, and often painful.

Climbing my way out of rock bottom wasn't just about stopping the drugs; it was about undoing the damage I had been inflicting since I was a teenager. I had spent years digging that hole, and now, filling it back in would take more effort than I ever imagined.

Once I started seeing progress, I thought about Mike. He had been struggling as well, and I knew that recovery would be just as challenging for him as it had been for me. So, I put his name on the waiting list at the clinic.

When his turn finally came, I drove him to Knoxville, just like Matt had done for me. We made those morning trips together, and with each drive, we grew closer. Maybe it was the twins in us finding our way back, surviving together the way we always had. Or maybe it was the silence of those rides that allowed our bond to reemerge, no longer drowned out by drugs or chaos.

Studies have shown that twins are far likelier to share addictions than standard siblings. A study published by Cambridge University in the journal *Twin Research and Human Genetics* in 2014 analyzed data from over 7,000 pairs of twins and found that the heritability of addiction was estimated to be around 50%, underscoring the role genetics can play. We had shared so much—our childhood, our struggles, and now, our fight for recovery. Thankfully, the same connection that made us susceptible to addiction seemed to work in our favor during recovery.

But even with Mike on his path, I realized that I had no blueprint for this version of myself. I had never experienced life as an adult without drugs. Sobriety didn't feel like freedom—it felt foreign, overwhelming. Life without the buffer of a substance to dull the edges seemed impossible to navigate. I didn't know who I was without the high, and every decision, every responsibility felt like uncharted territory.

So, I decided to take it one day at a time. And when a full day felt too overwhelming, I broke it down to hours—just one more hour without using. One more breath. One more chance. Slowly, I learned to exist in a world I was unfamiliar with, knowing that each step forward, no matter how small, was still progress.

CHAPTER 18

CLIMBING OUT

When I finally had my life back in order—or at least what felt like order—I decided it was time to get off methadone. My debts were paid, I had a steady job, and I was finally saving money. But the truth is, that decision didn't come from a place of careful thought or readiness; it came from anger.

I had gotten into an argument with my counselor at the clinic. She wasn't someone I respected—at least not back then. She had never been an addict herself. To me, she was just someone who went to college, read textbooks, and earned a degree in substance abuse counseling. I figured she knew more about addiction on paper than I ever would, but I was too stubborn to care. In my mind, she couldn't understand what it felt like to live through it.

The argument started over something small: weed. I still smoked occasionally because it helped me relax, and after everything I'd been through, I figured I *deserved* that much. But during one of our sessions, she pushed back, telling me I couldn't fully recover if I kept using any substances. We couldn't agree on anything that day.

Frustrated, angry, and tired of being told what to do, I snapped. "Fine," I said. "I'm done. Let's start my detox right now." She tried to warn me, explaining that detoxing from ninety milligrams of methadone would be dangerous—I didn't care. I wanted to get off, and I wanted to do it as fast as possible.

I asked her the fastest way they could legally detox me, and she replied, "Ten milligrams a week." At that moment, I thought it was the right decision. Looking back, I realize how much of it was driven by pride and anger.

The first few weeks of detox weren't as bad as I had expected. Reducing my dosage by ten milligrams at a time felt manageable. At first, I hardly noticed the difference, and with each reduction, I reassured myself that I had made the right choice. I felt as though I was winning the argument, as if the discomfort proved I had been right all along. *See,* I thought. *I can handle this.*

But methadone detox feels like a storm you believe you've outrun, only to find it's been closing in on you all along. What began as mild discomfort quickly morphed into a dull ache that followed me throughout the day. By the time I reached forty milligrams, the ache had transformed into waves of nausea, chills, and sudden sweating episodes. Striking memories of past withdrawals—the sleepless nights, muscle spasms, and bone-deep pain—began to creep back into my mind, like shadows lurking at the edges of my thoughts.

I tried to ignore the signs, attributing them to temporary side effects. However, each week, as the dose decreased, the symptoms intensified. I couldn't shake the anxiety creeping through my chest or the restless feeling that gripped my legs when I attempted to sleep. The false sense of victory I had felt in those early weeks was gone, replaced by a growing fear that I wasn't as strong as I had convinced myself I was. Then, on June 4, 2003, I took my last dose.

The first twenty-four hours without methadone are deceptively calm because of its long half-life. My body felt tired, but it wasn't anything I couldn't manage. I brushed it off as the normal fatigue of the rapid detox from ninety milligrams and reassured myself that I had everything under control. Maybe, just maybe, I had gotten off easy this time.

By the forty-eight-hour mark, that optimism had faded. The dull ache I had felt during the tapering process returned, but this time, it wasn't diminishing. A light sweat clung to my skin, and a familiar wave of unease settled over me.

I attempted to distract myself with the mundane—cleaning, pacing, mindlessly staring at the television—but nothing could alleviate the growing discomfort. The anxiety crept in slowly at first, and then all at once, transformed my thoughts into a chaotic mess. My mind raced, oscillating between irrational fears and vivid flashbacks of previous relapses.

By the third day, eating felt impossible. My stomach twisted violently at the thought of food, and even a sip of water triggered waves of nausea that left me breathless. My legs wouldn't stop moving, twitching as if they had a mind of their own. I tried lying down, but the pressure building in my joints and the crawling sensation under my skin made it unbearable. The restlessness was maddening, and the realization that I had never gone more than two days without methadone made everything worse. This was uncharted territory, and I had no idea how deep it would go.

By day four, I was drenched in sweat, alternating between freezing chills and scorching heat that felt like my body was being microwaved from the inside out. My mind was foggy and sluggish, but my body was in overdrive. Every breath felt shallow, as if I was fighting just to stay upright. Sleep was out of the question—each time I closed my eyes, it felt like I was falling into a pit, only to jolt awake seconds later with my heart racing.

By day five, I felt like I was ready to die. I couldn't sleep, couldn't eat, and couldn't find comfort in anything—whether I was sitting, standing, or lying down. My bed was drenched from cold sweats as my body struggled to regulate itself, leaving me shivering in the damp sheets. Sleep-deprived and exhausted, I only got up to stumble to the bathroom. Most nights, I would take a pillow and blanket with me and lie on the cold bathroom floor next to the toilet. The effort it took to walk back and forth between my bed and the bathroom felt insurmountable.

By now, the dull aches had transformed into full-blown agony, resonating deep within my bones. Every joint, muscle fiber, and nerve felt like they were being held to a flame, burning from the inside out.

Any movement sent shockwaves of pain through my body, but staying still was just as unbearable. My mind wouldn't rest. The anxiety gnawed at me, making even the act of lying down still feel like torture. And through it all, right on cue, stood my worst enemy: the Addict Brain. *Just take one pill. It'll help with the withdrawals—you don't have to go cold turkey! Just one. One night of sleep, that's all we need. Just take one, and this pain will go away. We're suffering. Why are you letting this happen?*

The Addict Brain kept whispering, promising relief and making the temptation to give in unbearable. I spent every moment fighting the pull to make the pain stop. While the physical agony consumed my body, it was the emotional weight—the loneliness, shame, and regret—that felt the heaviest. In those moments, the memories of the people I'd hurt came flooding back, but one memory and one person stood out more than any other: Mandy.

A few months before I quit the methadone clinic, Mandy moved to Myrtle Beach, South Carolina. She never knew I was on methadone because I was too ashamed to tell her. I had lied more than once, telling her I had quit doing OxyContin, so she thought I was clean. Facing the truth of my addiction was hard enough for me to bear—confessing it to the woman I loved felt impossible. I didn't want her to see the version of me I had become.

In the months leading up to her move, we barely saw each other. She despised the town we lived in and believed building a life somewhere new was the only answer. I wanted her to stay, but I was in no position to ask her to build a life with me. I was a mess—a complete shit-show. Deep down, I knew the truth would eventually come out, and she'd be disappointed all over again. So, I didn't resist. I did what any guy in the friend zone would do for the girl he loved—I helped her move.

Every time she came home, we spent time together. During those visits, it felt like nothing had changed, as if the distance didn't exist and we were just... us. It wasn't the life I wanted with her, but it was better than not having her at all. With each visit, I held onto the quiet hope that she'd say she was homesick and wanted to move back, but that day never came. The sad truth was, she was happy at the beach. She had found what she was looking for, and I was left behind.

By the second week of withdrawals, I felt as sick and depressed as I had ever been. I still hadn't slept, and every single fiber of my being ached. Trapped in my apartment, I was consumed by pain and isolation, with nothing but my own thoughts for company. I thought that maybe getting out of the house would help and that a change of scenery might offer some relief. Just as that thought crossed my mind, my phone rang.

It was Erin, a mutual friend of ours who worked with Mandy. In the morning, she was driving to Myrtle Beach to see her and asked if I wanted to come along. Call it divine intervention or fate, but whatever it was, it felt like exactly what I needed.

I didn't have the energy or the money to drive seven hours to Myrtle Beach, but tagging along with Erin felt like my lifeline. Neither Mandy nor Erin knew I had been going to the methadone clinic, and they had no idea I was in the middle of severe withdrawals. I thought maybe, just maybe, if I forced myself to act normal and hide the symptoms for the next few days, I could trick my brain into thinking I was fine. But the truth was, I was far from fine. But I was desperate to feel anything other than miserable.

The next morning, Erin pulled up to my apartment. Her windows were down, music blasting, and she had a fresh pack of cigarettes and a joint already lit. I opened the passenger door and slid into the seat. She didn't say a word—just handed me the joint. It was exactly what I needed: wind blowing on my face, music blaring, and a fresh high to dull the edge.

I knew it wasn't ideal to smoke weed or cigarettes while detoxing, but it wasn't an Oxy or methadone, so I didn't care. The high briefly muted the symptoms, and any strange behavior I couldn't suppress, I could easily blame on being stoned.

The drive with Erin felt like a breath of fresh air. We reminisced about the old days, laughing at stories of Mandy and the fun we all shared. It was the first time in weeks that I had a genuine conversation with someone who didn't mention drugs, detox, or my recovery. For those few hours, I felt something close to normal again.

When we arrived at the beach, seeing Mandy for the first time in months made me forget how crummy I felt—at least for a while. But as the weekend dragged on, my energy plummeted, and it took everything I had just to keep my head up and my eyes open. My dopamine levels were so depleted that even getting out of bed felt like a monumental task. I blamed my exhaustion on working late at the bar and the long drive, and if Mandy noticed anything off, she didn't mention it.

I planned to tell her everything eventually, but not that weekend. During those few days, I was simply happy to be near her, grateful for the distraction from the pain that still gripped me.

When it was time to leave, the ride home with Erin felt different from the ride there. The wind whipping through the open windows didn't sting as much, and the miles stretching ahead seemed manageable. I couldn't tell if it was because the first two weeks of sobriety were behind me or if the time spent with Mandy had softened the rough edges of my mind. Either way, something inside me had shifted. The cravings were still present, but for the first time, they didn't feel insurmountable. I could breathe just a little easier.

We talked more on the way back—about life, the future, and even small, meaningless things. That's what I needed: a normal conversation, the kind that wasn't bogged down by withdrawal. With every mile that passed, I felt a weight slowly lifting, as though some small part of me believed I could actually get through this.

When Erin pulled into the parking lot of my apartment and put the car in park, she turned toward me, expecting the usual small talk to close out the ride. But instead, I reached into my pocket, pulled out my pack of Camel Lights, and handed them to her.

She raised an eyebrow. "I thought you said you were going to quit *after* this pack."

I opened the pack and stared at the eight or nine cigarettes still inside. "If I'm going to quit, smoking eight more isn't going to matter," I said, handing them over with a finality that even surprised me.

She gave me a look that was somewhere between disbelief and support, but she took them, and that was the last time I ever smoked a cigarette.

Just like that, I had quit two of the hardest substances known to man—cold turkey. Cigarettes had been as much a crutch as any drug, and I had tried to quit them at least a hundred times before. The Addict Brain had always convinced me to buy *one more pack*, always promising that *this* would be the last one. But this time, there was no negotiation, no bargaining. I didn't even have a plan—I just knew that I was done.

Perhaps it was sheer willpower, or maybe it was the momentum from quitting everything else that pushed me over the edge. Either way, when I closed that car door and watched Erin drive away, I realized I wasn't just fighting addiction anymore—I was beginning to win. One small victory at a time, I was one step closer to reclaiming my life.

Even after thirty days, there was no escape from my mind—no break, no interruption of thought, and no way to shut it off. The mind controls everything: our heart rate, breathing, blood pressure, hand movements, emotions—everything. And mine had been hijacked by addiction for years. The drugs I fed it had burrowed deep into every corner and recess of my brain, like an unwanted guest that refused to leave.

Now, stripped of those drugs, my mind was working overtime, and the cravings were relentless. I tried to set boundaries, to tell the cravings to stop or to push through the pain, but the moment I let my guard down, the Addict Brain would seize control, begging me to give it what it desired.

Even sleep wasn't safe. When I did manage to drift off, I would dream of shooting OxyContin, the sensation so vivid that I woke up in a cold sweat, convinced I had given in. My mind was my own worst enemy. I couldn't hide from myself. Every second of existence felt like a battle of wills—me versus the monster I had created.

I was exhausted, broken, and trapped in a cycle of craving, guilt, and self-doubt. But I wasn't done yet. Not this time. This was my rock bottom, and as painful as it was, I knew it would be my first foothold as I started climbing out.

CHAPTER 19

NO ONE IS COMING

Two months into recovery, I was beginning to sleep better, but it still wasn't restful. I tossed and turned most nights, and the lack of quality sleep made it harder to heal mentally. Some nights felt endless, and the thought of taking a pill for one good night's rest was a tempting whisper in the back of my mind.

The constant fatigue, combined with the daunting realization of how much work was ahead of me, felt crushing at times. Yet, amidst that exhaustion, I learned the most important lesson: I put myself here—no one else. This lesson came at a steep cost, one only I could pay.

It was like a unique invoice made out to John Mathews, and whether I wanted to or not, I had to pay it back – in full. It would take both time and money to repay the debt, but in the process, I gained something else: wisdom. Wisdom that can never be bought; it can only be earned. It was a simple but hard truth: no one was coming to save me. If I wanted to get out of this, it was going to be up to me and me alone.

This was something I would have to learn over and over again, but with each experience, my self-confidence grew. It wasn't just the realization that help wasn't coming—it was the understanding that I didn't need it to. I had finally discovered the source of my ancestor's strength buried deep within, and I realized it had been there all along. I was enough. I didn't need someone to hold my hand or reassure me that everything would be okay because now I knew it would be. That belief, rooted in the hard work of recovery, became my foundation for the future.

Over time, my mind and body eventually healed, and the withdrawals faded into a distant memory, but the Addict Brain never went away. To this day, it still lives inside of me, waiting for me to let my guard down. Suggesting I could have "just one" without any consequences when a doctor prescribes me painkillers after surgery.

Even when I resist, that same recurring nightmare resurfaces. The one I've had for decades. The dream that I've relapsed. Convinced I've thrown everything away, I wake up drenched in sweat. In those moments, the panic feels as if I'm starting from scratch, fighting to reclaim my sobriety one day at a time.

It's a reminder that, first and foremost, the relationship I had damaged the most was the one from deep within myself. I might have spent years deceiving those around me, but the most devastating lie was the one I told myself—again and again. Each time my subconscious whispered lies to justify my behavior, I believed it. I clung to those lies because they offered temporary relief from the shame I felt, but over time, those lies chipped away at my sense of self. As it turns out, the worst betrayal wasn't what I had done to others—it was how deeply I had betrayed myself.

When you lose trust in yourself, you carry a burden you can never set down. We are all born with an instinct to protect ourselves, a built-in mechanism that shields us from harm. But when your own mind becomes the source of your destruction, that mechanism breaks. I could no longer trust my own emotions, decisions, or even my innate ability to protect myself.

It's the same feeling when a close friend or your partner lies to you. Only this time, you can't simply walk away or remove them from your life. That option is not on the table. I was trapped in a cycle of second-guessing, and I questioned every thought. The person I was supposed to rely on—me—had become the enemy.

I did my best to silence the voices, forgive the person I once was, and accept the choices I had made. Slowly, I was able to reassure the scared little boy inside me that everything would be okay. Regaining trust in myself wasn't easy; it was the hardest battle of all because, for so many years, I had let myself down and betrayed my own promises. Untangling myself from the shame of those memories and believing I could become someone different was a daily struggle.

As for all the other relationships I had damaged, they also needed amending. The people I had hurt were both family and friends. I had burned so many bridges and told countless lies. I had lost everyone's trust, and losing someone's trust is like emptying a bucket of water. It takes only moments to pour it out, but refilling that bucket is a slow, arduous process.

Trust doesn't flow back in steady streams; it returns one drop at a time. Each drop is earned: through a right decision, a kept promise, or an honest apology. And when you're refilling that bucket, you realize that with each drop, the level barely changes. Sometimes, days or even weeks go by without adding a single drop. To complicate matters, the people you betrayed must be willing to watch you refill that bucket, knowing it could take years—sometimes even decades—before it's full again.

Not everyone was willing to wait. Some people walked away for good, and I couldn't blame them. Watching them leave was painful, but it taught me that not everyone is meant to stay in your life forever. But those who stayed gave me more than just a second chance. They gave me time to prove myself. There were days when I felt like I had barely made progress, but their willingness to wait reminded me that even the smallest drops could eventually fill the bucket.

As I worked to rebuild trust with others, life around me continued moving forward, bringing new challenges and old wounds to the surface. During this period, Mom sent a letter to all three of us, explaining her decision to divorce Dad. To this day, Mike has never read the letter that Mom sent us. He was angry at her for a long time for leaving him, and while I understood his anger, I never held it against my mother. It was probably a long-overdue exit, one I imagine couldn't have been easy to make. She has suffered a lot of guilt over the choices she did or didn't make as a parent, but now, after raising a child of my own, I understand the complexities of parenthood. I know she did the best she could with the resources and strength she had at the time.

There's a certain level of grace you develop for your parents after becoming a parent yourself—when you're the one making tough decisions, knowing that every choice will inevitably shape your child. Your kids will see you through the lens of their experiences, and you can't control how they perceive you, no matter how much you wish you could. It's only later, when life humbles you, that you begin to see your parents' humanity in a new light.

Shortly after I received Mom's letter, my father came to visit me. He was at his own rock bottom. We hadn't talked much since the day he kicked me out of the house, yet here we were—standing on a level playing field for the first time in our lives.

When I invited him in, we both stood silently for a while, almost like strangers. The room felt heavy, as if neither of us knew where to start. Then, out of the silence, came the words I had longed to hear and never expected: "I'm sorry, John," he said. He didn't elaborate or try to explain himself, but I could tell how much it took for him to say those words. Without hesitation, I stepped forward and hugged him.

In that moment, something shifted. For years, I had carried deep-seated anger toward my father, but as I looked into his eyes, I noticed something I hadn't seen before—vulnerability. Pain. He had been carrying his own burdens all these years, much like I had carried mine. Forgiving him wasn't about letting him off the hook; it was about freeing myself from the resentment that had weighed me down for so long.

While forgiving my father and working to mend the other relationships in my life had brought me emotional peace, it didn't erase the physical toll my addiction had taken on my body. After years of neglect, the damage wasn't just mental; it was also visual. My teeth were rotting, my arms had scars, and my body looked malnourished. Poor hygiene and a lack of nutrients had waged war on my appearance.

I needed to get my diet back on track and gain the weight I desperately needed. I had withered away, depriving myself of proper nutrition. Regaining my strength would require more than just calories—it would require balance. That balance was found in the gym. Since I was no longer smoking cigarettes or doing drugs, running and resistance training became my new source of pleasure. Something I hadn't felt since I was a kid. Working out gave me the same rush I had once sought through pills, only this energy was sustainable and lasted all day. Slowly but surely, I began to rewire my reward system. My brain, which was once conditioned to crave OxyContin, started to recognize and seek the natural high from the endorphins pulsing through my body after a workout. Day by day, little by little, my brain learned to crave something healthier than the destruction it had once pursued.

But no matter how much weight I gained or how strong I felt, I couldn't escape the constant reminder of my past when I smiled. My teeth were awful, and every time I looked at my reflection in the mirror, I saw the man I no longer wanted to be. I hated my teeth. I covered my mouth when I spoke. I tried not to smile, but when I did, I would catch people glancing at my cavities. Their eyes would flicker for just a second, and in that brief moment, my shame consumed me. The judgment—whether real or imagined—was suffocating.

I knew I had to fix them if I wanted to regain my confidence. I didn't have insurance, and I certainly didn't have the money for major dental work, but I refused to let that be an excuse. Eventually, I found a dentist who agreed to help. He offered to fix one or two teeth at a time, warning me that some were so damaged the fillings might not last, but I didn't care. I wasn't looking for perfection.

After a good weekend of bartending, I would take my cash and make an appointment. The transformation wasn't overnight, but tooth by tooth, appointment by appointment, I rebuilt my smile. And with each repair, I could feel my confidence returning. The man in the mirror no longer had to hide behind shame. I could smile freely, laugh without covering my mouth, and face the world with my chin up and chest out.

I still had to sort out all of my financial problems and criminal records—both of which were disasters. I was a convicted criminal, and collection agencies hounded me so much that I considered changing my number. Over the years, I had maxed out credit cards with no intention of paying them back. When all the accounts went to collections, my credit score hit rock bottom, making the idea of buying a car or a house feel like a pipe dream.

So, one by one, I answered the calls from debt collectors, negotiating payment plans and trying to salvage what I could. To this day, I'm pretty sure I am still blacklisted from ever holding a Discover or Capital One card. This also wasn't a quick fix. It took me seven years of steady payments before those debts finally disappeared from my credit record. Seven long years to correct mistakes that only took a matter of months to make. I wouldn't have a decent credit score again until I was thirty years old.

Resolving all my legal issues also required time and money. To make matters worse, my driver's license was revoked, and getting caught driving on a suspended license was the last thing I needed. So, I prioritized its reinstatement – which was no easy task. I had to complete mandatory driving classes, pay reinstatement fees, and maintain high-risk SR-22 insurance for a minimum of five years.

All of this cost money, a lot of money. As if that wasn't difficult enough, if I were late or missed one insurance payment during those five years, they would suspend my license again, forcing me to start the process over – mandatory driving classes, reinstatement fees, and another five years of high-risk insurance. However, I didn't let that deter me; since no one was coming to save me, I stayed the course. I worked hard, paid my reinstatement fees and the SR-22 insurance every month for the next five years.

As for my criminal record, I faced it head-on. Unable to afford an attorney, I had to represent myself. It turned out to be more time-consuming than complicated, and the process took less than a year to complete. I traveled to each courthouse, filed a written petition stating my case, and requested that my charges be expunged. I was responsible for the court costs and any filing fees. None of these exceeded a few hundred dollars each, but when you're living on a bartender's salary, paying a dentist to fix your rotten teeth, making SR-22 payments, handling credit card repayments, and covering everyday expenses, it added up quickly. Money was tight—*beyond* tight—but I didn't give up. I stayed the course.

It took me years to piece all the fragments of my life back together. Throughout the process, there were moments of doubt, negative people at every corner, and times when I questioned whether any of it was worth it. But I never listened to the naysayers. I never gave up, and most importantly, I stayed the fucking course. I kept my head down and moved forward, making one right decision at a time, never knowing what the outcome would be.

To this day, I have never attended a Narcotics Anonymous (NA) meeting, and I wouldn't see a professional therapist until later in my life. I also don't know how many times I tried to quit or how many times I relapsed, but what I do know is that this last time worked. It was a messy process, and it didn't happen all at once, but for me, the MAT program saved my life. With a combination of sheer will and grit, I was able to get clean and stay clean.

I have come to realize that no one consciously decides to become an addict. The path to addiction is often long and winding. Influenced by numerous factors. Without recognizing the complexity of these factors, empathizing with their struggle can be challenging. While I can only share my own experiences, I hope my story offers insight into a condition that is frequently misunderstood and unfairly judged.

Addiction affects individuals regardless of their race, ethnicity, or socioeconomic status. It is critical to understand that addiction is not merely a choice but rather a chronic, relapsing brain disease that impacts and overtakes every aspect of the mind and body. It requires ongoing support and treatment, with its roots often stemming from a combination of genetic, environmental, and psychological factors. It's vital to try to understand an addict's story so you can approach them with compassion and empathy.

Understanding what it takes to get clean and live on this side of the line has given me the perspective to be non-judgmental toward others and offer grace. Recognizing the level of shame involved when you realize you've lost control of your mind yet continue to lie to yourself and those around you has provided me with a grace mindset. Shame is a deep, painful emotion that arises from a sense of failure and is often tied to how we perceive ourselves in the eyes of others. It's linked to feelings of worthlessness and regret over our actions. So, when you are constantly reminded of how shitty of a person you were while using drugs, that kind of shaming can feel unbearable.

For ten years, I numbed all the pain and childhood trauma with drugs. I dug a hole so deep that no one could come in to save me, and the only way to climb out of rock bottom was to rewire my entire brain. I had to carve out new pathways in the reward system so that the pleasure centers wouldn't depend on drugs for stimulation. By staying the course and through consistent, sustained effort, I rewired those pathways. I didn't realize it at the time, but I was engaging a part of my brain that had lifelong benefits.

The Anterior Midcingulate Cortex might seem like a complex term to you, but it is more significant than you realize. It's the sole source from which grit and the will to live originate. This part of the brain *only* develops when you do something you don't want to do. Not when you do something you *have* to do, but specifically when you engage in something you *do not* want to do. So, it makes sense that this area of the brain is larger in professional athletes and smaller in people who continually struggle to maintain a healthy weight growing when they successfully diet, but it's particularly enlarged in those who have overcome challenges like addiction. The best part is that it maintains its size.

Over time, with each right decision, I strengthened a part of my brain that many would give their lifesavings to possess, but as I mentioned earlier, this cannot be bought; it must be earned.

My brain wasn't my enemy anymore—it was my greatest ally. I unlocked a potential I didn't know existed. I had done the impossible—climbed my way out from rock bottom. I didn't just rebuild my life—I rebuilt myself. I felt like I could do anything, and that's precisely what I did.

CHAPTER 20

WHEN EVERYTHING CHANGES

I continued to visit Mandy while she lived in Myrtle Beach. In December of 2003, I went to see her for the weekend to get away and give her a Christmas gift. It had been six months since Erin and I had taken that road trip together, and it was the first time that I felt like myself in a long time. I had gained a healthy amount of weight and was slowly regaining my energy, but I was still relearning who I was without drugs. I was trying to navigate nearly every aspect of my daily life: how to hold a conversation, engage socially, and manage stress without the numbing effects of drugs. I was rediscovering the world, one awkward encounter at a time.

We had a wonderful weekend catching up and spending time with Mandy's friends between her shifts at the bar. In the morning, I was set to leave. I packed my things and sat on her bed, waiting while she finished getting ready for work. A few minutes later, she walked out of the bathroom with a distressed look on her face.

"John," she said, "I'm late."

I knew enough to know she wasn't talking about work anymore. My mind flashed back to a roadside rendezvous a few weeks earlier when she came back home for her birthday. On our way back from dinner, we pulled off on the side of the interstate and climbed into the back of my pickup truck. I'll leave the rest for your imagination.

I told her I would buy some pregnancy tests while she was at work. With reluctance, she turned towards the door to leave.

"No matter what, it's going to be okay," I said, trying to sound reassuring, though I knew there wasn't much comfort I could provide until we had an answer.

So, I drove to the drug store, picked up three different pregnancy tests, and placed them on the coffee table. As I sat there waiting for her to get home, I tried to convince myself that this wouldn't change everything—but deep down, I knew it already had.

Several hours later, Mandy walked through the door. Her eyes lingered on me for a moment before shifting to the boxes on the table. She walked over without saying a word, picked them up, and headed to the bathroom. Her roommate and I waited in the kitchen for what felt like forever.

When Mandy finally came out, she held up the tests and said, "Well, I'm pregnant."

All the color drained from her face as she shook her head and covered her mouth with her right hand, the weight of what she had just said hitting her all at once. She was living the life she wanted—a life away from East Tennessee. Now, everything had changed. She was visibly upset and vocally expressed her frustration towards me, claiming it was all my fault. What should have been a happy moment for most couples felt awkward and tense. Her roommate, having witnessed the entire scene, stood silently, leaving me unsure how to respond.

We were both just twenty-two years old. She was bartending, and I was barely six months clean. We were two kids dealing with our own issues—and now we had to figure out how to raise a child of our own. We weren't exactly the poster children for child-rearing at that moment in our lives.

I had no idea how we were going to do it. We didn't have a place to live, my car was barely running, we didn't have medical insurance, and on top of all that, I was broke. But once I got home, I sat down and wrote Mandy a letter. I wanted to reassure her that everything was going to be fine. I told her I would do everything in my power to provide the best life I could for our family. That I would do whatever it took, even if that meant working two jobs for the rest of my life. I wouldn't leave her or walk away from my responsibilities as a husband and father. We were going to make it work, no matter what, and we would do it together.

Years later, during our divorce, she sent that letter back to me. Reading it again felt crushing. I had fulfilled many of those promises, but I failed to uphold the vow to love her until death do us part. But, when I first wrote those words, I really meant them. That much will always be true.

We decided to wait until after Christmas to tell everyone about the baby. Mandy's parents are religious, so she was worried that we would ruin the holiday by revealing that she had become pregnant out of wedlock. When we finally broke the news, they were thrilled. We'd been together since high school, so they knew me, and they knew how much I loved Mandy.

Even still, I knew how important it was to them, and to Mandy, that we got married. I remember sitting in church next to Mandy's grandmother when she leaned over to me, pointed at her wedding ring, and then pointed at Mandy.

I bought a six-hundred-dollar ring from a local jeweler with a white gold band and a small diamond in the center. It was all I could afford at the time. I carried the ring in my pocket for weeks, waiting for the right moment.

I had already asked both her parents for their blessing, and her dad simply told me, "I know you will take good care of her," as he shook my hand.

Now, all that was left to do was to ask the question. So, one night, I went to Mandy's mom's house, where Mandy was living until the baby came. I sat down on the couch beside Mandy while her mom stepped out for a cigarette. I had all these ideas about how I would like to propose to her, but sitting on that old couch where we had sat hundreds of times felt like the right moment.

With the television on in the background, I bent down on one knee and asked Mandy to marry me. She looked at me confused, "Where did you get the money to buy that?" was all she said.

A few moments later, her mom returned to the living room and saw the ring Mandy had slipped onto her finger. Her mom turned to me and said, "Well, what did she say?" "Mandy hasn't answered me yet," I responded.

"He already knows the answer," she said as she fidgeted with the ring.

This wasn't quite the resounding "yes" I had hoped for, but it certainly wasn't a "no," either.

I now had a new purpose - my family. Determined not to disappoint them, I started to think seriously about a career. I was still working as a bartender and bringing in decent money, but I knew the lifestyle and the late hours wouldn't serve me well in the long run. I needed to find a more sustainable form of income.

My dad had been a firefighter, so I knew it offered the benefits and stability I needed to support my family. The next day, I thumbed through the phone book and called every Fire Department within thirty miles of Sevierville to find out what I needed to do to get a job. They all told me the same thing: I needed to return to school, become a certified Emergency Medical Technician (EMT), and then volunteer somewhere so I could get some experience.

That same week, I enrolled at Walters State Community College and applied to volunteer at the Gatlinburg Fire Department. The department never called me back, even after several attempts to meet with the fire chief. However, I was accepted into First Responders School, a prerequisite for the EMT program. So, at the start of the Spring 2004 semester, I was officially a college student.

The course consisted of a four-hour class, one night a week, and only lasted for a semester, which made it easy to keep my shifts at the bar. I also picked up another job working for a contractor building cabins. I needed to earn any extra money I could to pay for school, and so we could find a small apartment in Sevierville.

Life doesn't slow down for anyone. In less than a year, everything changed. I had quit two of the hardest substances known to mankind: opioids and nicotine. I found out I was going to be a father, proposed to my girlfriend, and enrolled in college while working two full-time jobs to make ends meet. It wasn't exactly what I had planned when I quit, but I felt happy, healthy, and excited about the direction my life was headed.

CHAPTER 21

AGAINST ALL ODDS

Mandy and I got married on Sunday, June 6th, 2004. We didn't have enough money for a big wedding, so we planned to have our pastor marry us in a small ceremony with just our families after the Sunday service.

But on Sunday, the pastor had a different plan. Midway through the morning sermon, he said to the congregation, "Hey, I know y'all came for a service, but who here would rather see a wedding?"

Hands started shooting up everywhere. It was clear we were not going to get our private ceremony. He called us to the front of the church. Mandy was almost eight months pregnant, and I was in a pair of old blue jeans. He performed the ceremony in front of the entire congregation.

After we had exchanged vows, he passed around the collect plate, and in place of an offering that morning, the congregation contributed to our new family. Our "wedding reception" was held across the street at the local Red Lobster, and as you can imagine, there was no time or money for a honeymoon. After the wedding, we returned home to our apartment, except now we were husband and wife.

Almost two months later, on July 29th, we drove to Sevier County Medical Center to have Mandy induced. With our baby bag in tow, we were eager to meet our little girl. However, that excitement would have to wait when our daughter made it clear she was not ready to make her entrance quite yet.

Mandy was in labor for thirty-six long hours, so we didn't get much sleep that night, and by the following afternoon, we were both exhausted. But when the doctor finally informed us that she was ready to deliver, all of that exhaustion turned back into excitement. I asked if I could assist with the delivery, and the doctor agreed. That afternoon, on July 30, 2004, I was the first person to welcome Ashlynn into this world. Wrapping my hands around my daughter for the first time was incredible.

However, that moment was short-lived. The delivery had seemed successful right up until the moment we realized the umbilical cord was wrapped around her neck, and she still had not taken her first breath. She wasn't crying, and she was blue from head to toe. The umbilical cord was starving her of oxygen. The doctor immediately came over and took her from me. She was rushed over to a nearby table while Mandy and I held our breath as we watched Ashlynn struggle to find her own. Moments later, her lungs let out a wail, and one of the nurses gave us an approving nod. She was okay.

A nurse finished checking the rest of her vitals and cleaned her up. She wrapped her in a pink blanket and handed her to Mandy. All three of us were crying. There have been a thousand words said and written about the experience of looking at your child for the very first time, but there is not a single word in the dictionary that defines what that moment feels like. My mind couldn't comprehend that this tiny, beautiful, new person was a product of her mother and me. It was the single greatest moment of my life, and it still is.

We brought Ashlynn home to our apartment to start our life as a family of three. I was newly married, one year sober, and now a father. Even though everything changed for me when Ashlynn entered this world, and I became responsible for something greater than myself, I still held a different perspective. As I looked down at this beautiful little girl, I realized that if I weren't clean, she would mean nothing to me. I wouldn't be present; I wouldn't have two jobs trying to make ends meet. I wouldn't be enrolled in college. I *would* be like every other deadbeat dad out there because my addiction would consume my every thought.

Addiction is the only feeling I have experienced that is more powerful than even love. Many want to believe that when my daughter was born, it would be all the incentive I needed to keep me from ever going back, but they would be wrong. If I was still under the control of my Addict Brain, even the deep, awe-inspiring love I felt for her wouldn't have been enough. I would have let her starve to feed my own addiction, just as countless mothers and fathers who are addicted do every single day around the world, and I knew that with every fiber of my being.

Addiction isn't something that lives in you; it's a part of you. It doesn't vanish once you stop feeding it drugs or alcohol; it simply takes on a new form. The truth was, and still is, that I will be an addict my entire life, even if I am no longer using. Understanding this, I had to find ways to channel my addiction. For me, that became my family and learning. For the first time in my life, I found myself motivated to succeed. I wanted to be a good father; someone my daughter would be proud of.

That mindset allowed me to continue pursuing my goal of being a firefighter. That same summer, when Ashlynn was born, I applied to EMT school, which was set to start in the Fall 2004 semester. This program was a little harder and longer than the First Responder's class. This one was a four-hour class, two nights a week, and lasted for two semesters. But I was motivated - I didn't just want to pass my courses; I wanted to be the best in the class. I wanted to prove to myself that I was worth something, that I wasn't the dumbass my father had made me believe I was all those years.

Since I still had two full-time jobs, things became a little harder to juggle over the next several months, but I stayed the course. Any spare moment I had was spent studying or with Ashlynn.

During this time, I never had a day off—not a weekend, not a weekday, not Thanksgiving, Christmas, or any holiday for that matter. I worked Monday through Sunday. My daytime job was construction, and my nighttime job was bartending.

During the week, I worked from 7 am to 11:30 pm every day. I would leave for work at 6 a.m. and work all day for a construction company. Then, I would get off at 3:30 p.m., go home, shower, and leave to go straight to class or to bartend.

EMT school was a night class for working adults on Tuesdays and Thursdays. Those two nights were the only two days I didn't bartend. On the weekends, during the day, I would study and spend time with my family, but at nighttime, I would have to leave to make money at the bar.

While I was busy doing everything I could to work, study, and squeeze in being a father, in the back of my mind, I knew that all of this could, in the end, be for nothing. After all, this was during a period when people looked at individuals doing drugs as criminals – they didn't have the understanding or knowledge they have today.

Whether you sold drugs or committed murder, it didn't matter; to them, it was all the same. As if that wasn't bad enough, there were naysayers all around me, both coworkers and family members (mainly my father), constantly telling me I would never have a career in emergency services or obtain a state paramedic license because of my past. Maybe they were right; maybe all of this would be for nothing, but I didn't pay much attention to the negativity, and instead, I just stayed the course.

As the last semester of school was coming to a close, I had to submit a petition to the state of Tennessee before I could qualify as an EMT due to my criminal record and history as a former drug addict. I had to provide all my arrest records and disclose my past opioid addiction. As an EMT, I would be a licensed medical practitioner, placing me in proximity to many of the same drugs I had abused for years.

I gathered letters from my professors. I provided character references and wrote my story in my own words. I shared my journey of becoming a father and expressed that I had never desired or pursued anything more in my life than I had this job. I communicated to whoever was on the receiving end of that letter that I wanted to help people, to do my part to make the world a better place. But still, I knew they might not take my word for it. Once I postmarked that manila envelope and sent it off, I knew that all I could do was wait and pray. In the meantime, I tried to manifest the possibility that it might all work out.

While I waited, I continued to work on my smile and expungements. I was still making trips to the local dentist, paying him cash to fix one cavity at a time while fitting in my court appearances throughout East Tennessee. I knew that even *if* my letter were approved, no one would hire me looking the way I did or after a background check on me.

So, no matter what happened, I would be ready both mentally and physically. Those months of waiting were perhaps the longest of my life. But then, finally, one night, I came home from work to find Mandy in the kitchen holding up a letter. I had been approved, and in May of 2005, I was officially a licensed EMT.

The timing was fortunate because a new fire department had just been built in Oak Ridge, about fifty miles northwest of Sevierville. This created some rare job openings among the other local departments in the region, and many firefighters left their departments to work in Oak Ridge. Eager to work anywhere, I applied for positions at the Knoxville and Gatlinburg Fire Departments. Both approved my application and scheduled me for their physical agility tests and written exams.

Knoxville never called, but Gatlinburg invited me to an interview. So, I bought my first suit and practiced answering questions they might ask me. When I walked into the Gatlinburg Fire Department for my interview, I was taken through the front door and back to a large conference room. Six men sat at a long table in front of a single black chair in the middle of the room. I introduced myself and sat down.

I took a deep breath and tried hard to stay calm. The fire chief, Gary West, started the interview with introductions. Each panel member had a list of questions, and they took turns asking them one by one. With each question, I gained more confidence. Up to that point, I felt like I was nailing the interview.

That was, until one of the men asked, "Mr. Mathews, we will be running a background check on each candidate. Will there be anything we find when we run yours?"

I felt like someone had just punched me in the gut. I had gone through the process of getting each of my offenses expunged, so I was doubtful my record would come up. But at that moment, I decided to take a gamble instead.

"Sir," I said, "As a matter of fact, you might."

They all immediately stopped writing and looked directly at me. That second, I felt every one of their heartbeats elevate, a trait I haven't felt so strongly since I lived with my father.

Then one of the men chimed in, "Son, what in the world are you doing here?"

Before I could answer, someone else asked, "Well, what exactly are we going to find?"

I then proceeded to give a condensed version of my rap sheet. I decided to be open about who I was and how hard I had worked just to get into this room. I shared a short version of my life story, leaving out an important part – the fact that I was a recovering addict. I wanted to be honest when answering his question; I wasn't, however, ready to be completely transparent. Remember, the world viewed addiction differently back then, and I wasn't ready to share that portion of my life yet.

I said, "Look, I know I have given you every reason *not* to hire me, and if I don't get the job, I will understand. During that time in my life, I was a different man – much different than the one that sits in front of you today. Yes, I made a lot of bad decisions, but I also understand no one made those decisions for me: they were mine and mine alone. So, if this is as far as I get, I will have to live with that because I made this bed, and I will lie in it, whether I like it or not. But I can promise you one thing: if you just give me a chance, I promise I won't let you down."

When the interview ended and everyone started getting up to leave, the oldest gentleman in the room, a retired chief, walked over to me and extended his hand to mine.

"Son," he said, "Honestly, I don't know if you will get this job or not, but what I do know is – it's not going to matter. I have sat through hundreds of interviews, but never one like that, so I wanted to shake your hand and say thank you. I have no doubt that no matter what you do, you will succeed."

His words meant more to me than the interview itself. It felt as if he was validating my very existence. It was the first time I had been able to share my thoughts with conviction and be recognized, hell, even *respected.* But as I left the room, I expected I would never hear from Chief West again.

But a week later, I received a phone call. It was the Gatlinburg Fire Department training officer, Captain Roger Ogle. "Mr. Mathews," he said, "We would like to formally offer you a job in our department as a firefighter/EMT."

I stood silently for what felt like minutes, trying to make sure I had heard him correctly.

"Mr. Mathews, did you hear what I just said?"

I steadied my voice and said, “Mr. Ogle, it would be an honor to accept the position.”

The moment I hung up the phone, everything around me seemed to fade into the background. I made it. I had clawed my way to that moment, and I fucking made it. I had climbed out of the hole I had dug for myself, fought through every doubt, every setback, and now I had something no one could take away from me. A chance. A real shot at being the man, the father, and the provider I had fought so hard to become.

I immediately fell to my knees and burst into tears. They were tears of both relief and disbelief. I wasn’t supposed to make it this far—not according to the statistics, not according to the people who had written me off, and certainly not according to my ex-firefighter father who said I would never have a good-paying job because no one would hire a former drug addict with a criminal record. But I had proven all of them wrong.

For the first time in my life, I had something stronger than the pull of my Addict Brain. I had a purpose. I had a family. And I had proof that no matter how deep you sink, it’s possible to climb back up, one step at a time.

There was no grand celebration, no moment to pause and catch my breath. Life kept moving forward, and I had to keep pace. But as I stood in that tiny apartment, looking at Mandy and Ashlynn—at the life we were building despite the odds—I understood something that would carry me through every challenge ahead. It wasn’t about the failures. It wasn’t about the past. It was about the choice I made every single day to push forward. To block out the noise. To stay the course. And for the first time, I believed with everything in me—I was going to make it.

CHAPTER 22

ROOKIE

A couple of weeks after receiving that call, I started my job with the Gatlinburg Fire Department. I was an unusual hire because I technically wasn't a certified firefighter. More specifically, I hadn't completed "Rookie School" at the Tennessee Fire Service and Codes Enforcement Academy in Bell Buckle, Tennessee. Remember when I mentioned I tried to volunteer with the fire department, but the chief never called me back? Well, now that same chief had just hired me, and I still didn't have any experience. So, part of my hiring agreement was that I would need to pass the state exam successfully, but at this point, I wasn't worried.

My first year in the fire department was largely spent in school learning about fire behavior. On each shift, I was assigned to work in the ambulance only as an EMT. Gatlinburg is one of nine municipal fire departments in Tennessee recognized as a combo department, providing both fire and EMS services. So, I essentially had a permanent seat in the ambulance until I graduated from the fire academy. If we had a structure fire, the only role I was qualified for was to be the team sherpa, also known as – the rookie.

However, I kept pushing and eventually completed my fire training courses, making me ready for the "Live Burn" at the state academy in Bell Buckle. The fire academy was situated in the middle of nowhere and resembled a military base. The sleeping quarters were cramped and uncomfortable, and everywhere you went, the smell of wet ashes lingered.

During my time there, I met some of the most incredible men and women. There's something about being in a burning building together that bonds you for life. We enjoyed some of the evolutions, while others tested our limits. In the end, most of us graduated, but a few didn't make the cut and left feeling defeated.

After spending a year in the department, I decided to pursue my paramedic license. I wanted to provide patients with the highest level of pre-hospital care available, and it would also increase my salary.

Paramedic School was significantly harder than the other programs I had to take to become an EMT. It was competitive and very challenging, and most students who began the program never finished.

The program consisted of forty-eight credit hours, but I felt like it took years off my life to complete. Even after I finished the college courses, I still had to take the National Registry Exam to obtain a state license, which is, to date, the hardest exam I have ever taken.

I managed to pass the test, and once I finished the program and became a paramedic, I was just a few credit hours shy of earning an associate's degree in pre-hospital care. I wanted a college degree, so I made an appointment with my counselor.

She explained to me that, based on my ACT score, I would have to take every remedial English class the college offered. I was horrible in English. Do you remember my English IV class, the "SAVE WATER, DRINK BEER" one? Well, now I was being told I had to take, not one, but *three* separate remedial classes before I could even take my first accredited college English 101 class. That meant it would take me five semesters (or two-and-a-half years of school) to obtain the English credits I needed to graduate. To put that in perspective, a normal college student could do it in two semesters or one year of school.

But I stayed the course and went on to earn my degree, graduating at the top of my class, summa cum laude. I even received an academic achievement award when I graduated for obtaining the highest GPA ever in the history of the paramedic program at Walter State.

During school, I became fascinated by human science and how the body functions. So, after graduating, I continued my education as a biology major. At the time, I aspired to be an anesthesiologist, so I took as many biology and anatomy classes as possible.

Looking back, I realize my desire to become a doctor stemmed from my ego. I believed it would be the ultimate way to showcase my comeback from being a junkie. Thankfully, that dream was short-lived, but the knowledge I gained in those biology and anatomy courses proved invaluable as a paramedic. Understanding how the body functions in theory was one thing, but witnessing it fail in real time was another.

My time at the fire department taught me about the fragility of human life. When you arrive at the scene of an accident or an emergency, it's not like a movie or an episode of *Grey's Anatomy*—you are confronted with real-life trauma. It's not natural to witness death.

Watching someone die is horrifying, and no matter how often you see it, it never becomes easier. I witnessed countless family members crying over the bodies of their loved ones. Seeing them grieve was difficult, especially when they had to watch as their loved ones took their last breaths.

I sat with mothers and fathers who had just lost every single member of their family in a fatal car crash. I responded to incidents where people were burned alive while trapped in wrecked cars. I stood over lifeless bodies at murder and suicide scenes and pulled mangled victims from twisted metal—many so disfigured, you could only tell their gender by their body parts. All of this was difficult and unnatural for the human mind to comprehend.

Too often, as emergency responders, we had to learn to accept things that were out of our control. No matter how bad the day was or how difficult the previous call was, you still had to show up and perform your duties at the next scene. No one cared if you had gotten in a fight with your spouse that morning or didn't sleep well the night before; they just wanted assistance—and not just any assistance, but the best care you could provide. And that is exactly what we did every single day.

Fortunately, there were more good days than bad during my time at the fire department. Most of the calls were routine, but some stood out and are seared into my memory.

Gatlinburg is a small town. Paramedics serve neighbors, friends, and sometimes even their own families. One morning, just after 7 a.m., our first call came in. Dispatch informed us of a pediatric cardiac arrest. We jumped in the truck, activated the lights and sirens, and headed out.

As we approached the home, the surroundings started to feel familiar. It was a friend's house, though I hadn't seen him since high school. Knowing what the dispatch said was happening in that house, I prayed to God that he had moved.

When we arrived, a woman rushed outside and began frantically pointing us toward the house. "She's not breathing, she's not breathing," she repeated.

We entered the kitchen and noticed a man on his knees, leaning over an infant and trying desperately to revive her. When I started asking questions, the man turned around and called me by my high school nickname, a name I'd earned for rolling blunts like a Cuban cigar.

"Cuban?" he asked, "Is that you?" With one glance, I knew it was my high school friend. My heart sank. "Oh, thank God," he said, "Please help me. Please help."

We did everything we could to bring her back to life, but when we arrived at the hospital, there was nothing more that could be done. Losing a patient is always hard, especially when it's a child. It's even more devastating when you know the family. I knew the second I laid eyes on that infant that she was gone. Still, I knelt beside my friend and went through the motions of trying to bring this child back to life. I persisted longer than protocol allowed, dreading every second until I would have to tell my old friend that his daughter was dead.

Working in emergency services helped me become skilled at compartmentalizing my thoughts, which made it easier to leave the stress of work at work.

At home, Mandy and I were working together to build a family and a future. For the first two years of Ashlynn's life, Mandy stayed home with her while I worked and attended school.

Money was tight, and for much of that time, Mandy and I felt like ships passing in the night, swapping parenting duties and squeezing in family meals whenever we could. We didn't have the luxury of family outings or trips to the zoo during those early years, but our home was filled with love and happiness. Mandy took on much of the parenting burden early on, yet she remained supportive of my goals.

Our plan was that once I obtained my paramedic license, Mandy would return to school to become a registered nurse. That had always been her dream, and it would be my turn to support her career.

After I finished school, Mandy started bartending again to earn some extra money and get accustomed to working once more. Eventually, she returned to school and became an RN.

Our parents helped us with Ashlynn constantly. They watched her whenever Mandy and I had scheduling conflicts, both day and night. After nearly three years, we had finally established a routine. Ashlynn made it easy to be a young father. She was happy, curious, and full of joy, and we did everything we could to make sure she always felt loved and cared for.

Our marriage had never been traditional or normal. We didn't experience the honeymoon years of being young and childless. We scrambled and hustled to make ends meet and support our little family. We devoted every ounce of ourselves to our daughter, but often, that left us with little room or energy for the time and effort that a marriage requires.

Mandy and I knew each other like we knew the back of our own hands. I knew her habits, idiosyncrasies, what brought her joy, and what triggered her wounds. We grew up together and made our way through the world as young teenagers, in and out of love, and later as partners and parents. Mandy was my best friend, my other half. But there's a saying that "what you nurture will grow," so it stands to reason that what you don't nurture will not grow either.

This process doesn't happen overnight. Just like a plant deprived of water won't die immediately, but over time, its leaves will gradually wither and shift from green to yellow. The roots will dry out, and that plant, once bursting with life and possibilities, will fade and shrivel onto itself.

CHAPTER 23

EARNED, NOT GIVEN

While Mandy was enrolled in nursing school, I continued to quench my thirst for knowledge. I enrolled in every course available to me, regardless of whether it was online or in-person.

I earned hundreds of certificates; some were specific to fire and EMS; others enhanced my leadership skills or aided me in complex rescue operations like swift water and high-angle rescues. I even became an instructor for the courses I enjoyed the most, teaching doctors and emergency personnel advanced life support. I never slowed down; I didn't have the time. I had wasted so much of my life using drugs, and now I needed to catch up.

I loved being a firefighter/paramedic, but I wanted more. I had an insatiable desire to challenge myself. Perhaps this desire stemmed from my early days of getting and staying clean and recognizing that if I could do that, what could I *not* do?

Leadership seemed to come naturally to me, so I set high standards for my goals. I wanted to be a fire chief and lead a fire department. It didn't necessarily have to be the Gatlinburg Fire Department; I just wanted the ability to grow and shape my own department.

To achieve this, I knew I had to distinguish myself from everyone else. I needed to build my resume so impressively that it appeared stronger than my colleagues' because I had to compensate for my fewer years of service. After all, I wasn't going to sit around and wait thirty years to become fire chief.

So, I separated myself from the group and studied every chance I had while still having a little fun. There is plenty of horseplay happening at almost every fire station in the United States. After all, it's a group of grown men and women stuck at a station for twenty-four hours at a time. So, when boredom hits, you start doing dumb shit. So, I want to be clear: I was *heavily involved* in these shenanigans, probably the chief instigator in many of them. However, when it was time to study, anyone on my shift would tell you I isolated myself and focused on my homework.

I was advancing my academic career with a bachelor's degree in business management and leadership, and I eventually obtained a master's degree in business administration. However, a college degree wouldn't be enough to set me apart from the other applicants. I needed more. So, I did something different—something nobody else was doing. I became a certified fire inspector.

Few took this class unless they wanted to become an actual fire inspector because the test was incredibly challenging.

I even failed my first attempt, but I gained a better understanding of the fire code, and I figured if I was going to be the fire chief enforcing the code, then I better know it.

A few years into my time at the fire department, I learned that the director of the Sevier County Emergency Management Agency (SCEMA) was set to retire. A friend of mine, David Hamilton, worked there and mentioned they were searching for someone to fill the position. So, he encouraged me to throw my name into the mix.

The Emergency Management Office is a subsidiary of the Federal Emergency Management Agency (FEMA) and is responsible for several important areas, including but not limited to training, federal and state grants, and managing large-scale incidents and disasters.

At that time, I wasn't sure it was what I wanted to do. However, I agreed to work at EMA part-time to see if I enjoyed it. So, for the next several months, I was back working two jobs again.

I enjoyed working at SCEMA with David. It presented a new set of challenges, and I felt part of something bigger than just the fire department. It wasn't long before the director stepped down, and the position needed to be filled.

David was the assistant director then, but he didn't want to be promoted. He liked what he did and felt comfortable staying in his current position, so he wanted me to apply.

I was excited about the opportunity. No, it wasn't the fire chief's position I had previously hoped for, but it was an opportunity to shape and run my own emergency service department. David was an excellent co-worker, and we had great ideas for growing the department. So, I applied for the position.

Since the Director position was an appointment, I wasn't sure I would get the job. I also knew that my past could be an issue. I didn't want any secrets or surprises if I were chosen for the position, so I asked a close friend of mine, Greg, if he would meet with me.

Greg and I met at the fire department. He came in as the fire chief after Gary West left. We quickly bonded, and he took me under his wing. He was one of the first men in my life to believe in me. Without Greg, many opportunities in my life would have never been possible. He is a great man and a dear friend, and I will always appreciate what he did for me.

We decided to meet for breakfast at the local Cracker Barrel and sat at a corner table. I had never shared my story with anyone before, not in its entirety, and I felt nervous. I was afraid he might see me differently or suggest that I withdraw my name for the director position. It felt as if I had been keeping something from him all along, and the weight of it left me feeling awful.

We started with some pleasantries, and then I dove in. "Look," I told him, "I am honored to have a chance at this job, but before we move forward, I need you to know a few things about me."

He sat a little straighter in his chair as I continued, "And I want you to hear it from me before you hear it from anyone else."

Greg listened intently as I shared my story. It was painful yet cathartic. By the end, I had a pile of crumpled-up Cracker Barrel napkins in front of me. I didn't expect to become so emotional, but it was the first time I had been transparent about who I truly was. I was no longer just answering questions honestly – I was now completely transparent.

Greg sat quietly for a moment before he said, "John, nothing you just told me makes me think any differently of you." My shoulders relaxed a bit, and I felt a wave of relief wash over me.

"I can't believe you've overcome so much and become the person you are today," Greg said. "You have an incredible story." "But don't tell anyone else," He quickly added. "This needs to remain between us. People won't understand."

My relief was short-lived, but Greg was right. People during that time would not have understood, and they would have falsely judged me. You must remember this was at the height of the opioid crisis, and the public viewed anyone involved as a criminal. Worst, they wanted all criminals locked up. Jails were just starting to become overcrowded because of the crisis.

It wouldn't be until years later that the justice system recognized this as a mental health crisis rather than a crime epidemic. Greg and I both understood this. He wanted to protect me because, even though he knew I was a good man with great character, those who didn't know me would have tried to destroy me. So, that was the last day I ever shared my story with anyone – until now.

In September 2010, at the age of twenty-nine, I was sworn in as the director of the Sevier County Emergency Management Agency. I felt a deep sense of pride, not only because of the title—though "Director Mathews" had a nice ring to it—but also because of the journey that brought me here. Just seven years ago, I had been at rock bottom. I was physically broken, mentally exhausted, and trapped in addiction. I had nothing: no direction, no stability, and no future in sight.

Yet, against all odds, I rebuilt my life from the ground up. Through perseverance, I pursued education, regained trust, and earned my position as a leader in emergency management. What I had accomplished was nothing short of extraordinary. I had transformed from a drug addict devoid of hope to a highly educated department head responsible for leading an entire agency. This wasn't just a career milestone; it was proof that resilience, determination, and the refusal to give up can completely transform a life.

Over the next several years, David and I grew the agency. We held multiagency training events. We managed several federal and state grants, allowing both fire and police departments to have better training and state-of-the-art gear.

I, of course, continued to learn and grow in this field as well. I had the privilege of serving on multiple boards, both county and statewide, many of which I chaired. Some of my proudest accomplishments included being appointed to the National Justice and Public Safety Policy Steering Committee for NACO and being elected as president of the Emergency Management Association of Tennessee. I attended meetings at the State Capitol and sat in the Speaker of the House's private chamber, socializing and sharing laughs with elected officials.

But often, I felt like a stranger. I still carried deep-rooted shame, knowing that if they discovered my past, they would view me differently. Whether they would or not, I didn’t want to find out, so I kept my secret. I constantly thought about how, a decade ago, a guy like me wouldn't have received a second glance from any of these people, let alone their attention.

No one warns you that when you finally step into the room where it all happens, you’ll feel like an imposter. It’s as if you think someone made a mistake or that it’s all a big joke, with you as the punchline.

Trying to fit into this new world, I watched YouTube videos about etiquette and table manners. No one had ever taught me which side of the plate the fork should be placed on or the proper way to cut a steak. Whenever I attended a fancy dinner, I carefully observed how everyone else set their glass down or arranged their cutlery on their plate after finishing. I didn’t want them to see me as some guy who just crawled out of the Florida swap or a hillbilly with no table manners. I desperately wanted to fit in.

I didn't know it then, but soon, those same people would look to me and my education, which went beyond fancy dinners and table manners. I learned there is another side to success, one that would test and challenge everything I believed about myself, along with the accountability and immense responsibility that come with leadership.

Because on November 23, 2016, two teenage boys hiked up the Chimneys trail in the Great Smoky Mountains, lit a match, and set an entire city on fire.

CHAPTER 24

LOVE LOST, LESSONS LEARNED

As my career began to take off, my marriage was slowly unraveling. When I stood in that church on my wedding day and promised Mandy that I would love her until "death do us part," I believed every word.

The promise I made to her didn't break all at once. It came apart gradually over the years through small hurts, little injuries, exchanged words, and things left unsaid. Each one created a crack that eventually couldn't bear the weight anymore, causing the whole thing to come crashing down.

We had been married for nine years and raised a beautiful daughter together, but I had drifted further and further away from the feelings of love I had experienced early on in our relationship. I convinced myself to stay for the sake of my commitment and to shield my daughter from the wounds of a broken family, but it seemed that the harder I tried, the more resentful I grew. Mandy was a good mother and was still my best friend, but she didn't feel like my partner anymore.

It would be nice to be able to point to one thing and say, "*There*, that's the reason I left my marriage," but logic doesn't apply when it comes to falling in love, and it certainly doesn't apply when falling out of love, either. I gradually fell out of love over time, and as I did, I moved closer and closer to that line I never thought I would cross. Until I finally did.

I have learned that the moment you say, "I would never do that," you're walking a tightrope. I have been that guy. I've drawn my own lines in the sand, believing there are boundaries I would never cross or actions I would never take.

I could make a million excuses or find ways to justify my actions, but the truth remains the same. I cheated on Mandy, and no matter how I viewed it, I was in the wrong. I chose the cowardly way out instead of being honest and telling her the truth. It seemed easier to give her a reason to leave, which felt less painful than ending the marriage.

I used to think there were three sides to every story: your side, their side, and the truth. But the truth seems much grayer than it used to. There are many ways to interpret our choices and the reasons behind our decisions, and there isn't always a clear narrative. Our choices are often filtered through the lens of our past experiences, traumas, and mistakes, limiting our decision-making to our own point of view.

I often wonder about the effect our divorce has had, or will ultimately have, on our daughter. I'll never forget the day I told her that her mother and I were getting a divorce, and she'll never forget it either. It's seared into both our memories, and it tore me apart to shatter her reality.

When she was younger, anytime we watched a movie or a commercial, there was always an image of a 'whole' family—parents together with their kids—and I often wondered what went through her mind.

Her version of 'normal' looks very different from what the world portrays. I've done my best to create a sense of home for her—a place where she feels she truly belongs. This isn't the life I once envisioned, and I imagine it's not what she pictured either, but I hope she has always felt nothing but love and a deep sense of worth, despite all my shortcomings as her father. She is the best thing I have ever done, and being her dad is the greatest privilege of my life.

We always hope that, somehow, the next generation will be better than we are. If there is just one lesson my past mistakes can prevent her from learning, I'll feel like I've done my job. I want her to grow and learn. I recognize that her own lessons will teach her some of that, but maybe, just maybe, she'll learn something from my failures and successes.

CHAPTER 25

FOR DAVID

After my divorce, I threw myself into work and focused on building a new support network. Many of those connections came from the colleagues I worked with most closely—one of whom was David. He was my right-hand man, the go-to person for every incident in our office, and, more importantly, one of my closest friends. Working side by side, we developed a strong bond, and I came to rely heavily on both his expertise and his friendship.

One night, David and I were responding to a wildfire—one of the hundreds we managed each year. Depending on its size, containment could take hours or even days. During those long nights, we often talked for hours, and on this particular evening, our conversation turned to our families.

I'll never forget the moment he looked at me and said, "If anything happens to us, I want you to know I'll take care of your daughter as if she were my own, and I hope you'd do the same for mine."

That night, we made a promise—almost as if we both sensed something was coming.

Less than a year later, David was gone. On Father's Day, he took his daughters to the Biltmore Estate in Asheville, North Carolina. Following his doctor's orders, he stayed up all night in preparation for a "daytime" sleep study scheduled for Monday morning. But when he arrived at the sleep center, the staff determined he was too exhausted for the test, as the results wouldn't be accurate. Concerned about his condition, they directed him to a room where he could rest before driving home. He never woke up.

Since then, I have made an effort to uphold my end of the bargain for David's two daughters. When David passed away, the words from the stranger who had assisted me on the side of the highway all those years ago came rushing back to me. I had promised him that I would pay it forward, and while I can't fill the void left by their father, I can be present and remind them of how much he loved them.

I struggled to come to terms with David's death; I was grieving the loss of my friend. No one in the office could bring themselves to take the photos from his desk or clear out his office, so every day, we walked past the empty room where he used to sit. We all felt his absence.

Meanwhile, his position remained unfilled. I thought I'd hire someone when things slowed down and the time felt right, but replacing him would mean further acknowledging that he was gone and wouldn't be coming back.

Somehow though, the world kept spinning, and we witnessed no shortage of natural and manmade disasters that year. A tragic helicopter crash claimed five lives. There were two suicides, including one where a man jumped from the Gatlinburg Space Needle, landing on the sidewalk in front of a group of tourists. A series of storms swept through the area, causing tornadoes and, of course, the extreme drought that plagued the entire state and caused hundreds of wildfires.

Wildfires are common in our area due to the high number of visitors each year. The Great Smoky Mountains National Park is the most visited national park in the country, attracting over fourteen million visitors annually. People come from all over the United States, and many are not familiar with the landscape or fire safety practices, making it common to encounter numerous fires caused by cigarette butts, improper fireplace use, and campfire negligence.

Until 2016, the most significant wildfire that David and I managed was the Black Bear Cub Fire in March of 2013. This fire was ignited by a hot tub on the back deck of a cabin. Due to the terrain and the proximity of the houses, fifty-three cabins were destroyed, and another twenty were damaged in the area.

At that time, it represented the most substantial residential loss from a wildland fire in Tennessee. Following large incidents, our team would return to the drawing board to figure out how we could improve for the future. We both enjoyed the problem-solving aspect of emergency management.

Each case felt like a new puzzle, presenting new pieces and complexities every time.

By the end of 2016, Tennessee's drought was getting worse. Due to dry conditions and a series of wildfires sweeping across the state, the Tennessee Emergency Management Agency declared a State of Emergency. Yet somehow, Sevier County had managed to survive the drought without a large wildfire despite the influx of tourists that year. That was about to change, and David's absence would be felt more than ever.

CHAPTER 26

THE GATLINBURG WILDFIRES

Monday morning, November 28, 2016, started like any other day. Thanksgiving leftovers were being tossed out, and Christmas lights were starting to go up. People woke up to head to work, parents rushed to get their kids off to school, and tourists packed up after a long holiday weekend.

For weeks, a smoky haze had hung in the air—a lingering reminder of the wildfires burning across the state. By now, it had blended into the backdrop, more of a nuisance than a real threat. But that morning, something felt different.

When people stepped outside, they weren't met with the usual distant haze. The air was thick and heavy in a way that made breathing feel unnatural. A dense smoke covered the town, clinging to buildings and creeping into every open space. Cars weren't just dusted with the typical fine layer of soot—they were coated in ash. The familiar had turned foreign. It was the kind of morning that sent an eerie warning, one that most didn't recognize until it was too late.

It all began on Wednesday, November 23, 2016, the day before Thanksgiving, when a fire ignited atop Chimney Tops, a popular hiking trail in the Great Smoky Mountains National Park (GSMNP). With over eight hundred miles of trails, the park was packed with visitors enjoying the crisp autumn air and stunning mountain views.

Hikers commonly document their treks with cameras or GoPros, and on this particular day, one hiker's footage revealed something far more unsettling. As he reached the summit, his camera captured two teenage boys striking matches one after the other, tossing them to the ground, eager to see if the dry forest floor would ignite. Most burned out on contact, but then, one did not.

The National Park Service responded to the fire as it always did—internally, without outside assistance. Initially, it appeared manageable, covering only about an acre and a half. However, Chimney Tops is no ordinary terrain. The steep, rugged slopes made access nearly impossible for firefighters, limiting their ability to contain the flames. In most wildfire responses, crews create a fire line by clearing vegetation and exposing bare soil to stop the fire's advance. But here, in this unforgiving landscape, cutting a fire line was out of the question.

The park's fire management team initially attempted to contain the fire within a containment zone using an indirect attack strategy. In this method, firefighters allowed the fire to burn to pre-determined control points rather than directly engaging it. By Saturday, November 26th, the fire had grown to about eight acres but was still within the containment area.

Conditions dramatically changed on Sunday, November 27th. Forecasts had suggested the fire would continue its slow downhill spread, but exceedingly dry conditions and rising winds changed everything.

That morning, relative humidity dropped, and wind speeds, although initially low, gradually increased throughout the day. By the afternoon, helicopters were deployed for water drops, though their capabilities were limited. By nightfall, the fire had expanded to thirty-five acres. Yet, no one knew what to expect the following day.

Late Monday morning, just before noon, personnel from GSMNP discovered a new fire near the Twin Creeks Science and Education Center, approximately a mile and a half from the Gatlinburg city limits. Shortly after, the Gatlinburg Fire Department (GFD) was dispatched at the park's request for additional units for structural protection in the Twin Creeks area.

At this stage, our office arrived on the scene at the GFD's request and followed the standard wildfire response protocol. We were there to help monitor the fire's movement and assist where possible. Our primary role was to coordinate shelter locations and facilitate mutual aid requests through state emergency resources.

The Sevier County Wildland Fire Task Force was activated, and by noon, the Gatlinburg Police Department

(GPD) began door-to-door voluntary evacuations in the Mynatt Park neighborhood, the closest residential area to the fire. Some residents heeded the warning immediately, while others hesitated because they were familiar with wildfires.

At 1:30 p.m., we set up the Gatlinburg Community Center as an evacuation shelter and informed the Red Cross to start mobilization efforts. That afternoon, I met my team at the command post, where we analyzed the fire's projected movement using a 3D topographic fire model. We incorporated real-time weather data, which allowed the software to estimate the burn rate. The software also considered the forecasted rain, which would help slow the fire's advance.

But that's not what happened.

Monday evening, November 28, 2016. After our 5 p.m. press conference, we drove up the mountain to see the fire for ourselves before darkness settled in. The fire was now about a mile and a half from the Mynatt Park community. As we navigated the winding road, I pulled out my phone and began recording. The video captured an eerie stillness—the soft crackling of embers, the deep orange glow flickering in the encroaching darkness. It was quiet. Too quiet. The timestamp on my screen read 5:50 p.m.

As I exited the park and headed back towards the command post, the first gust of wind hit my truck and pushed it sideways, rattled by a sudden, powerful wind. I rolled down my window—papers flew across the cab, nearly sucked into the night. I pulled over next to the emergency crews stationed along the road and stepped out to speak with them.

The wind nearly ripped my truck door off its hinges. Just minutes earlier, the air had been still. Now, it felt like something else entirely. What we didn't know then was that a cold front was sweeping through, driving out the desperately needed rain we had been counting on.

Storm radar recorded Category 1 hurricane wind gusts reaching eighty-seven mph, a force strong enough to hurl embers miles ahead of the main fire line. The once slow-burning fire had turned into a full-blown firestorm, an unstoppable inferno racing toward Gatlinburg.

Later, this event would be named "A Mountain Wave Event." A phenomenon rarely seen in the Smokies, where high-velocity winds surge down the slopes. At the firestorm's peak, between 6 p.m. and 11 p.m., the wildfires accelerated at an unprecedented rate of over 2,000 acres per hour. That is over half an acre every second. This fire wasn't coming. It was already here.

For generations, the people of the Smoky Mountains have lived with the threat of wildfires. They have witnessed dozens, perhaps even hundreds, come and go, mostly contained and mostly controlled. However, this would be a fire unlike anything anyone in the Smoky Mountains had ever experienced in their lifetime.

Back at the command post, the scene was organized chaos. Inside the Emergency Operations Center (EOC), phones rang relentlessly, radios crackled with updates, and a growing sense of urgency filled the air.

Calls were coming in from every corner of the city, reporting new fires. We dispatched every available crew to each incident and deployed firefighters to evacuate the most threatened areas in a systematic manner. *But the calls kept coming.*

And soon, we faced the unthinkable: we ran out of firefighters. There were simply too many fires and not enough personnel to battle them all.

At 8:14 p.m., the city lost power. Soon after, a fire destroyed two major cell towers, cutting off all mobile communication. The fire hydrants lost water pressure as pump stations failed due to the outages. Gatlinburg was now in the dark—both literally and figuratively. Shortly after 8:30 p.m., a total evacuation order was issued for the city. From that point on, all communication had to take place via radio.

Just before my cell service was completely cut out, I managed to make one last call to the Tennessee Emergency Management Agency (TEMA). We wanted them to issue an evacuation alert using the emergency broadcast system, which would send it to all mobile devices, similar to the Amber Alerts we receive. But just after I relayed the message, the line went dead.

Later, we learned that TEMA attempted to call me back to confirm the message, but I never received the call. The evacuation alert never went out. Even if they had managed to send the emergency message, it likely wouldn't have reached anyone's cell phone due to the loss of service and power outages.

The communication outage made commanding the scene even more challenging, and worse, it contributed to the delay of the public evacuation alert. With cell service down and emergency channels overwhelmed, getting critical information out became nearly impossible. Then, something unexpected happened.

The National Weather Service, which had never before issued an evacuation alert for a wildfire, managed to make independent contact with our E-911 office and sent one out.

At 9:03 p.m., they sent an alert via the Emergency Alert System. [xiii] This is the same system you see when watching television or listening to the radio. You may recognize the voice: "We interrupt this broadcast to bring you an important message."

"The City of Gatlinburg and nearby communities are being evacuated due to wildfires. Nobody is allowed into the city at this time. If you are currently in Gatlinburg and are able to evacuate...evacuate immediately and follow any instructions from emergency officials. If you are not instructed to evacuate...please stay off the roads." [xiii]

It wasn't perfect. It wasn't the message we had planned. But at that moment, it was enough to save lives. Thanks to the National Weather Service message, along with the efforts of local news media and social media, at least 14,000 people were able to escape safely.

The wind speeds continued to escalate with each passing minute. Inside the fire station, the bay doors rattled violently, and the glass panels shuddered under the pressure. It became clear that the station was no longer a safe location. Chief Greg Miller and I decided to relocate the command post to the Gatlinburg Community Center, the same location where evacuees had already been sent.

When I tried to exit through the front door, the wind's pressure was so intense that it took all my strength to push it open. As soon as I stepped outside, the wind nearly ripped the door from my hands. I ran to my truck and headed toward the Community Center.

When we finally arrived at the Community Center, it was complete chaos. Hundreds of people crowded near the front entrance—families with small children, their worldly possessions stuffed into backpacks that had once held nothing more than schoolbooks and packed lunches. More evacuees arrived by the minute, their faces a mix of confusion, fear, and exhaustion.

By 2 a.m., the winds had calmed, the rain had finally arrived, and the radio traffic had slowed.[xiii] But the damage was already done. In just six hours, the fire consumed 17,000 acres and destroyed over 2,500 structures. To put that into perspective, a single house fire usually requires at least two fire engines, each containing four firefighters. That meant, under normal conditions, it would take more than 5,000 fire engines and 20,000 firefighters to respond to each incident. We didn't have those kinds of resources.

Listening to the radio that night was gut-wrenching, with panicked voices pleading for help. The men and women responding to those calls looked defeated, exhaustion etched on their faces as they gathered in the Community Center parking lot. Some had streaks of tears mixed with ash, wiping their faces as they came to terms with the reality that their own homes had likely burned.

Early that next morning, as we drove through the city, it looked like a war zone. Power lines laid tangled across the streets. Hurricane-force winds had ripped trees from the ground and toppled light poles, leaving roads completely impassable. Stop signs had melted into a puddle of metal. Vehicles still smoldered, their aluminum wheels liquefied by the heat, and their tires disintegrated, leaving them resting on bare frames.

The main parkway in Gatlinburg, the same street where I had once been arrested all those years ago, was now silent and empty. No tourists filled the sidewalks. No neon lights lit up the night. No music drifted from bars and storefronts. Only the crackling of fire and the piercing wail of fire alarms echoed through the streets.

As we climbed higher up the mountain roads, my mind was fixated on the people who lived there. I had driven these streets hundreds of times, but now they were unrecognizable. No mailboxes. No road signs. No indication that homes had ever stood here. Only concrete slabs and charred, hollowed-out vehicles remained. It wasn't just devastation—it was erasure.

Many of these families had lived through generations of wildfires, never fearing for their safety. They had seen flames before and watched them flicker on distant ridges. But this fire was different. It disrupted everything—every belief, sense of security, and previously held notion of danger. It destroyed everything and everyone in its path.

CHAPTER 27

MANAGING THE AFTERMATH

As the number of evacuees increased, the Community Center became overwhelmed. It simply couldn't operate as both a command post and an evacuation shelter, so we relocated the evacuees across the street to Rocky Top Sports Park, an 86,000-square-foot athletic facility that would now serve as a refuge for thousands.

At that time, Lori Moore, the facility director—and a dear friend—took charge of shelter operations. Without her, it wouldn't have run as smoothly as it did. She coordinated with the Red Cross and volunteer organizations, managing the 3,000 people who stayed there for weeks.

They organized logistics for food and housing donations, provided a full-service kitchen serving three hot meals a day, and even set up a temporary pharmacy to refill life-saving prescriptions. A clothing station was created where families could "shop" for donated clothes, selecting what little they needed to move forward.

By Tuesday morning, the walls at the shelter entrance were covered with photos—faces of missing loved ones and pets, their names written in desperate pleas. Bulletin boards that once displayed sports schedules and community events were now filled with prayers and last-known locations. The Red Cross launched a reunification effort, but days and weeks later, the number of missing remained in the hundreds.

While Lori and her team concentrated on the shelter, we focused on the missing people and the 2,500 burned structures. Search and rescue teams were sent home by home, looking for the lost, the trapped, and the dead. But with no road signs, mailboxes, or recognizable landmarks, the search efforts were severely hindered. Entire neighborhoods had been erased. And with cell towers down, GPS was useless.

We went back to the basics: printed maps, hand-drawn grids, and manual counting systems. As teams completed each search, they spray-painted a large "X" next to the structure, marking the date, time, hazards present, and the number of victims found. It was slow, exhausting, and devastating work.

Inside the newly organized command post, leaders from across the state and community collaborated to solve problems, delegate tasks, and allocate resources. During briefings, officials from every department gathered in the same room, coordinating their responses together. Each briefing concluded with solutions rather than excuses. Not every mission went smoothly, and the loss was overwhelming, but some days, a miracle would happen, and it would re-ignite our efforts again, reminding us that there were people out there counting on us not just every day, but every second.

At Westgate Resort, a timeshare community, multiple cabins had burned down. A couple who was evacuating the resort found themselves trapped in an elevator when the power went out. Using the elevator phone, they were able to call 911, and a crew was dispatched; however, by the time rescuers arrived, the buildings were lost. Everything was reduced to smoldering rubble. There was no way to locate the victims.

Hours passed, and the couple called 911 again. Weakened by smoke inhalation and disoriented, clinging to their last hope, they had one final message: "Tell our family we love them." Dispatchers were stunned—they were still alive.

A rescue team was immediately redeployed to search the wreckage for any sign of where the couple might be. Suddenly, a faint voice responded to their calls. The team rushed to an intact elevator shaft, pried open the doors, and pulled the couple to safety. Miraculously, they survived.

There were dozens of stories like theirs—people who had narrowly escaped, families separated and later reunited, pets found days later, alive and waiting. Yet, even as hope flickered in some hearts, others continued to wait—still holding on, still searching, still praying for the impossible.

The command post was constantly buzzing with activity. Each day, we held two critical briefing sessions—one in the morning and one in the evening. These meetings ensured a smooth transition between twelve-hour operational shifts, provided vital updates to the media, and kept the public informed.

As the operational periods continued, the hours turned into days. We tirelessly searched for victims, cleared roads, fed responders, and assessed homes and properties. The work was as emotionally draining as it was physically exhausting, but we had to compile every piece of data that was coming in from the field. Not only were thousands of people still waiting to hear if their homes were standing, but we also needed this data to submit a formal aid request to FEMA.

The solution came through geographical information systems (GIS). Led by Stacey Whaley and her team, they developed a large-scale reporting system that both emergency personnel and the public could access. It wasn't easy. Her team sifted through thousands of handwritten reports, meticulously entering data throughout the night to map the destruction.

By Thursday morning, the GIS map went live, allowing people to search their addresses and see if their homes had survived. While this was devastating news for many, it provided clarity—something so many families desperately needed.

Every day, the rescue crews would return after another long day of searching, and we would gather in a small office after the operational briefing to get the latest information on their search and rescue efforts.

That night, Chief Miller stood before us, delivering the latest update, "As of this moment, we have found and rescued seventy-four individuals, who are all being treated for their injuries."

He then took a deep breath and paused before continuing, "We can also confirm four new deaths of individuals who were found and recovered today. That brings our total loss of life to eleven souls." Heads dropped. No one spoke.

Whenever a victim was discovered, rescue teams secured the scene and called in law enforcement to document the findings. Each body was then transported to a mobile morgue set up at the base of the mountain by the Knox County Forensic Center. The fire complicated identification, compelling forensic teams to take their time to ensure every victim was identified properly and their loved ones notified with dignity.

After the briefing ended that night, I stepped into a small office just off the large gathering room. I needed to be alone—to catch my breath and process my thoughts.

The following morning, I would face the press and the public for the first time. My name and face will forever be tied to the 2016 Gatlinburg Wildfire. Everything I did or didn't do, said or didn't say, would be documented, reviewed, and scrutinized—both publicly and privately.

I stayed in that room, allowing my thoughts to slow, gather, and collect themselves. When I finally glanced at my phone again, it read 11:43 p.m. I knew I needed sleep, but my mind remained restless.

Stepping outside, I walked to my truck and looked toward the mountains, their shadowed outlines resting against the faint glow of the moon. For the first time in two days, the sky looked calm, and the thick scent of smoke was fading. Nature had already begun the healing process. But for this community—for all of us, including me—the road to healing was just beginning.

CHAPTER 28

THE PODIUM

On Thursday morning, December 1, 2016, the media gathered outside the Anna Porter Public Library, waiting for an update on the wildfire. The community, the region, and the nation wanted answers.

How did this happen? Who was responsible? People needed something—or someone—to blame. I knew some of those pointed fingers would land on me.

Until now, much of what transpired in the days and moments leading up to November 28th remained largely unknown to the public. This was our moment for transparency—our attempt to explain what, in many ways, felt impossible to articulate.

Even those of us who had been on the ground floor were still sorting through the facts. Every day, every hour, we were learning more about how and why this disaster had unfolded the way it did.

That morning, I walked from the command post to the library. Through the window, I could see the microphones and cameras separating me from them. I made my way to the back room to meet with Chief Miller, Tennessee Governor Bill Haslam, and other local officials. On the other side of the wall, I could hear the reporters clamoring through the doors. I excused myself to use the bathroom.

When I closed the stall door, I felt the urge to throw up, but instead, I leaned my head against the cold metal, staring down at my feet. In that moment, I wasn't John Mathews, Emergency Management Director. I was twenty years old again—broken, helpless, a stain on society. I was a fraud. *Who did I think I was?* I had been here before. Maybe not in this exact stall, not in this exact building, but I had stood in this same spot—doubting, fearing, questioning.

Only this time, I knew that everything I had been through had led me here for a reason. I put my face in my hands and whispered the same words that had haunted me for years:

"How the fuck did I get here?" Would they see through me? I thought. *Had they dug into my past and planned to attack my character? Was I about to be exposed, judged, or ridiculed?*

I straightened my shirt, took a deep breath, splashed cold water on my face, and looked up at the mirror. This wasn't where I thought I would be; that much was certain. But everything I had been through until this moment—every failure, every rock bottom moment, every second of self-doubt—hadn't disqualified me. It had prepared me. I walked down the long hallway, my footsteps echoing on the tile. I pushed through the last door and headed toward the podium.

The room fell silent, but the lights shining down on my face felt like they were shouting at me. I stepped up to the microphone, and immediately, hands shot up in every direction. I pointed at the first hand I saw.

"Mr. Mathews, when asked whether the evacuation alerts had gone out, you responded 'yes,' when, in fact, the texts never actually went out to the public."

I steadied my voice and said, "If people did not receive the message we sent out, of course, we are unsatisfied with it."

The truth was, and still is, that learning those text messages never reached Sevier County residents devastated me. I understood that, ultimately, the responsibility started and ended with me. That was part of the job—I had signed up for it. Still, I felt crushed.

Yet, I was grateful the National Weather Service had taken it upon itself to send an evacuation alert to TV and radio stations, even if their service didn't include text message alerts. Many would argue that the text message would not have gone out, as the cellphone towers were already destroyed.

I continued answering questions, speaking only to the facts I knew at the time. One reporter pressed me on evacuations, comparing our fire to one in Asheville, North Carolina, where no one had died because of an early evacuation order. But those two fires weren't the same.

The Asheville fire was slow-moving, while ours had hurricane-force winds. Months later, we received a letter from the North Carolina Fire Chiefs Association apologizing for the reporter's comments and defending our actions. They understood what the public didn't—these two fires were *not* alike.

More information would unfold in the days and weeks ahead, but that day, the public learned the death toll had risen to eleven people. They heard their names. Who they were. And for the first time, they also learned that the fire had likely been caused by human hands—a fact that, until that moment, had been known only to EMA, the fire department, and the National Park Service.

This was, and still is, one of the hardest truths to accept: that all the death and destruction that had occurred over the last several days was no longer just the result of a natural disaster gone wrong; it was preempted and expedited by some*one*, not something.

CHAPTER 29

MOUNTAIN TOUGH

The 2016 Gatlinburg wildfire was one of the costliest and most devastating wildfires in U.S. history, with an estimated total exceeding one billion dollars.

It constituted the largest mobilization of resources in Tennessee's history and marked the first time a non-coastal community in the U.S. evacuated an entire town and kept it closed for five consecutive days. A total of 225 agencies from across the country responded with 445 fire trucks, 3,500 responders[xiv], 3,000 individuals sheltered[xv], 20,000 volunteers, over 16,000 meals served[xvi], and $12 million donated.[xvii]

The resources and donations had poured in. The Rocky Top parking lot became the staging area, home to more than one hundred firetrucks from the surrounding region, who had volunteered their resources to help contain the fire. The parking lot also housed a steady flow of semi-trucks full of donated items from the surrounding area and beyond, wanting to help the community.

The Pigeon Forge Rotary Club operated Boyd's Bears distribution center in Pigeon Forge, which distributed food, clothing, furniture, medication, and other items to thousands of fire victims. More than three hundred volunteers staffed the center, and daily shuttles ran from Rocky Top so that fire victims housed at the shelter could go and retrieve any necessities they needed.

In the days and weeks following the fire, we launched a website called Mountain Tough, where individuals could go to find help or offer help. A multi-agency resource center (MARC) was established at an old, abandoned mall, where people could obtain new documents lost in the fire: driver's licenses, social security cards, and marriage certificates.

The way the Gatlinburg and surrounding communities came together to support one another was unlike anything I had ever witnessed. People showed up day in and day out, taking care of their neighbors. They fed them, clothed them, and made sure they knew they weren't alone.

The Zac Brown Band even came and prepared food for all the first responders and those at the shelter. Out of their goodwill and pockets, they served 5,000 people that day.

But when the ashes settled, and the fire was no longer the main news, our community was still hurting. The effects of the fire continued to linger. Today, you can drive through the city of Gatlinburg and look up to see the tree skeletons on the mountain where the fire carved its path. Some buildings and homes were never rebuilt, and entire communities were decimated and remain empty. There are individuals still healing from the trauma of fleeing and displacement.

The final fire victim was identified two months after the fire broke out. The victim had been missing since Monday, November 28, 2016, and was the last confirmed death from that tragic day, bringing the total death toll to fourteen.

Eventually, the donations stopped coming in, and the shelter closed its doors. People tried to return to their lives, piecing together what had been lost, but few were the same. I certainly wasn't. Life didn't return to normal for me—there was no going back to who I was before the fire. Every negative comment, every critical remark on social media felt like a personal attack. And often, it was.

The two teenage boys who allegedly started the fire by striking matches on the Chimney Tops Trailhead were charged with aggravated arson. In Tennessee, Aggravated Arson is a Class A felony, the most serious type of felony.

However, in 2019, prosecutors dropped the charges due to the lack of evidence showing they were liable for the fire. The judge noted that while the boys' actions were "immature and thoughtless," there was no evidence that they intended to start the fire or that their actions were reckless enough to warrant legal liability.[xviii]

The public reacted with mixed feelings. Some believed the boys should be held accountable for their actions. Others felt the lawsuit unfairly targeted them, making them scapegoats for a tragedy that was largely beyond their control. Some also argued that the focus should be on preventing similar fires in the future rather than assigning blame for the past.[xix]

The Gatlinburg wildfires sparked extensive public debate over the complex issues surrounding wildfire prevention and response. Many commended the heroic actions of the individuals and first responders who worked tirelessly for days and weeks. Others were quick to criticize park officials and government agencies, arguing that a stronger wildfire prevention plan could have mitigated the disaster.

But one truth remains: everyone involved—responders, government officials, and emergency personnel—did an outstanding job. The men and women who filled the command center for days without rest were fearless. Together, we tackled impossible challenges with the limited information and resources we had at the time. No one—no expert, no official, no outsider—could have done better. No matter what the armchair quarterbacks said, we knew the truth.

In the wake of the tragedy, we focused on progress. While we couldn't go back and change what had happened, we could channel our energy into ensuring that residents and visitors were better prepared for future emergencies. Over the next two years, local officials and emergency responders worked to enhance communication systems and evacuation protocols.

We hired a third party to compile an after-action review to identify areas for improvement, and many of those recommendations were implemented. Yet the reality remains:

disasters are, by definition, beyond control. No amount of preparation can prevent every tragedy. And when the next crisis comes—because it always does—it won't be about what could have been done differently in the past. It will be about who shows up, who stands together, and who refuses to let devastation define them.

Gatlinburg did not burn and fade away. It endured. It rebuilt. And for those of us who lived through it, the fire is more than a memory—it's a reminder of resilience, of loss, of the fragility of everything we take for granted. But above all, it's a testament to the unbreakable spirit of a community that refused to be consumed by the flames.

CHAPTER 30

PTSD

The fires humbled me in ways that still shape who I am today. At the time, I didn't realize it, but I had developed Post-Traumatic Stress Disorder (PTSD) from the wildfires. I had kept myself so busy—moving from one decision to the next, one crisis to another—that I never stopped long enough to process what I had been through. Work was easier to navigate than my own emotions, and, if I'm being honest, I was always better at solving other people's problems than dealing with my own.

During the wildfires, my role allowed me to lean into my greatest strength: problem-solving. Years of challenges and training had sharpened that skill, and in the heat of the crisis, I put it to the test. In the aftermath, that ability made me highly sought after.

Headhunters reached out with offers—some from regional organizations, others from out of state. I gave some serious thought to a few of them. But in the end, this was my home. This was my community. And I knew there was nowhere else I was meant to be.

I was eventually promoted to assistant mayor of Emergency Services by Mayor Larry Waters, who had become my friend and mentor over the years. He was, and still is, someone I greatly admire and respect. However, as proud as I was to be promoted, I received a job offer that would change my life. It would allow me to stay in the area but take me out of the public sector and into the private sector. I wrestled with the decision to leave the mayor's office. I didn't want to let people down—people like Mayor Waters, who supported me, believed in me, and, most importantly, allowed me to succeed.

I spent months wrestling with the decision, but deep down, I knew I had to leave in order to grow. I could have stayed in that job forever—I knew the work inside and out and understood every expectation—but that was exactly the problem.

Some people seek that kind of comforting stability in a career. I had climbed the ladder of success in emergency services, even surpassing my early aspirations of becoming fire chief, but I needed a new challenge. It was time to move forward.

The decision was not easy. I had several conversations with Mayor Waters leading up to my decision to leave local government. I valued his opinion. He became more than just a friend and mentor; he became the father I had never had. Telling him I had accepted the job offer was not easy. In fact, that conversation is still one of the hardest I have ever had. But

Larry supported me in every way. I left, knowing I had done the right thing, but that didn't mean it didn't hurt.

After leaving, I wrote Mayor Waters and Gatlinburg City Manager Cindy Ogle a letter expressing my deep gratitude. Each letter was written with profound love and attention, each unique to them. I tried my best to explain what their support has meant to me and how they both gave me a chance when many people wouldn't have.

A month later, I started my new job as the vice president of operations for D&S Builders, a local building company. D&S Builders is an extremely successful company led by Alex and Mary Davis, with gross annual revenue exceeding $150 million. I was honored to be a part of their team, and over the years, I learned so much from Alex and his team.

As I adapted to my new role and faced a new set of challenges and problems, I encountered something I hadn't expected—silence. I had left a job where I was the go-to person, the one with all the answers, the problem solver in every crisis. Now, I found myself in unfamiliar territory, where I knew almost nothing.

Looking back, I realize there was something addictive about always having the answers—about being needed, about the constant urgency. The chaos had given me purpose, and in some ways, the busyness itself had been a kind of high. Without it, I felt unmoored, forced to confront the stillness I had long avoided.

For the first time in almost a year, my phone wasn't constantly pinging. The hum of urgency that had defined my existence was gone, leaving behind a void I didn't know how to fill. The first few weeks felt like freedom. The first few months felt like suffocation. Without the adrenaline of crisis management, I found myself jumping at noises that weren't there, waking up gasping from dreams I couldn't remember, and feeling the weight of something I refused to name pressing against my chest.

When the constant stimulation I had experienced for the better half of the previous year slowed down, all the emotions I had walled off from feeling came to the surface in a powerful and all-consuming wave. The first of which was depression. In the hustle and bustle, I had failed to recognize that my mental health had been in a state of decline for longer than I likely knew.

There wasn't necessarily a direct source or a throughline that I could pinpoint and trace as the culprit for my sadness, but the weight of the feeling was overwhelming.

Then came the panic attacks.

I experienced my first one when I woke up in the middle of the night in a cold sweat. My heart wasn't just pounding—it felt like it was trying to escape my chest. My hands were clammy, my breaths shallow, and for a moment, I couldn't tell if I was suffocating or just losing my grip on reality.

The people in my life who know me best would likely describe me as an upbeat, positive, and energetic person. While that person others see is the real me, I struggled in ways that nobody knew and didn't know how to talk about.

I lacked the tools to communicate how I felt and how to seek help. I thought back to my father and the way he allowed his depression to consume him, branching out like the roots of a tree, growing new tendrils underneath the surface, and becoming bigger and bigger over time.

At that point in my life, I didn't believe in depression, or at least I didn't believe *I* had depression; that was something reserved for other people, not me.

I kept telling myself things like, "John, snap out of it," 'Get over it,' "You're better than this," and "Get your shit together, man." But my body wasn't listening.

No amount of willpower could make my heart stop racing. No pep talk could lift the invisible weight pressing against my chest. I had outworked every challenge in my life—why the hell couldn't I outwork this? And yet, the more I tried to force it away, the deeper it sank its claws into me.

This shouldn't come as a surprise, but according to the data, men are less likely than women to seek treatment for mental health conditions. In 2019, the National Survey on Drug Use and Health found that among adults with any mental illness, women were 11.2% more likely to seek out mental health treatment than men.[xx]

Additionally, men are more likely than women to die by suicide. According to the American Foundation for Suicide Prevention, men die by suicide at a rate that is 3.63 times higher than women.[xxi] Men also account for most of the suicides in the United States; in 2019, men died by suicide at a rate of 22.3 per 100,000, compared to a rate of 6.2 per 100,000 for women.

According to research, men's help-seeking behavior is influenced by a complex interplay of factors, including gender role socialization, stigma, and masculine norms that prioritize self-reliance and emotional control.[xxii]

In other words, I felt like less of a man, and I didn't want to admit it to anyone, including myself.

However, I continued reminding myself that I had already overcome the most challenging mental hurdle: getting clean. That thought helped me push through and get my mind to a place where I could eventually find ways to move through the depression in healthy ways. It was like something finally clicked in my brain and reminded me that I was in control of how I was feeling and that I could push past the normal limits of pain because I recognized that much of it lived inside my head.

I had to stop punishing myself for feeling the way I did and start focusing on how to move through it. Change had to start small. I couldn't think about fixing everything at once—I just needed a foothold.

So, I made two choices. First, I started meditating, hoping to bring some order to the chaos in my mind. I downloaded an app, and every morning, I sat in silence with my coffee, listening to a guided meditation. At first, it felt pointless. My mind raced, my body fidgeted, and I questioned why I was even trying. But I stuck with it.

The second thing I did was to start running. I remembered how good it had felt in the days of early recovery when I pushed my body forward and forced my mind to follow. I dug through my closet, found an old pair of Nike sneakers, and threw on the closest thing to running shorts I could find. I grabbed some shitty headphones, laced up, and took off. That morning, I barely made it half a mile before my lungs burned and my legs begged me to stop. But the next day, I woke up and did it again. And then again. One mile turned into two. Then five.

Still, I kept going. I upgraded my shoes, got better running gear, and even downloaded a training program. I found a few friends who ran, too, and we pushed each other. Ten months later, I ran the Knoxville marathon. The same body that once betrayed me in addiction, the same mind that had tried to convince me I wasn't strong enough to keep going, had carried me across a finish line after running 26.2 miles – one I never thought I'd reach.

Running and meditation reminded me of something I had first discovered in recovery—the incredible elasticity of our brains. We are so quick to assume our limitations are set, that we are who we are, and that change is impossible. But I had already proven otherwise. I had seen it firsthand when I fought through withdrawal and clawed my way out of rock bottom.

And now, here I was again, in a different battle but armed with the same weapon: my mind's ability to adapt.

The power of neuroplasticity is something most people never truly witness in themselves. Yet, I had the rare opportunity to experience it twice. The first time, it saved me my life from addiction; the second, it pulled me out of depression as a top executive – both at each end of the spectrum. Each time, I was forced to rewire my brain to break old patterns and replace them with new ones. And through that process, I gained something invaluable—wisdom. I understood, in a way that no textbook or expert could ever teach, that our brains are capable of profound transformation if we are willing to put in the work.

It's a truth backed by science. Studies have shown that practicing a new skill or engaging in a new activity can lead to changes in the structure and function of the brain, as well as the formation of new neural pathways. For example, learning to play a musical instrument, speak a new language, or memorize a poem can all lead to changes in the brain's neural pathways. [xxiii] Certain lifestyle factors, such as exercise, diet, and sleep, can also influence neuroplasticity and promote the growth of new neural connections.[xxiv]

But science alone isn't enough—you have to live it. I had to learn the hard way that there's no shortcut to rewiring your mind. It takes effort, repetition, and a relentless commitment to doing the work. Running and meditation didn't magically fix my problems, but they provided the tools I needed to shift the narrative in my head. They allowed me to harness the very thing that had once been my enemy—my own mind—and use it to heal.

No, I didn't get better all at once. And even now, it's a daily choice. Healing isn't a finish line you cross—it's a commitment you make over and over again. But the most important thing you can do is put yourself in the best possible position for success.

At the end of the day, no matter how different our stories are, we're all just human beings learning as we go. My path may not look like yours, but at our core, we all wrestle with the same emotions: happiness, sadness, fear, pleasure, anger, guilt, shame, regret, and more. We stumble, we fall, and we get back up. And more often than not, we must learn the same lessons over and over before they finally sink in. Growth doesn't happen on a schedule, and transformation isn't about perfection—it's about persistence.

For most of my life, I had to hit my own breaking point before I was ready to change. It wasn't enough for someone to tell me I needed to do better—I had to *feel* the cost of staying the same. And I don't think I'm unique in that.

We all have moments when we're simply not ready, when the pain hasn't outweighed the comfort of the familiar. And that's okay. The key isn't beating yourself up for not being ready—it's recognizing that when the time comes, you will be. And when you are, that's when everything shifts.

No one can force you out of a toxic relationship, a dead-end job, or destructive choices—not until you're ready to leave. No one can tell you when your moment of clarity will arrive. But trust this: it will come. And when it does, when you've reached the point where staying the same is no longer an option, that's when the real work begins.

So, if you're struggling right now, if you're caught in a cycle you don't know how to break, give yourself grace. But be honest. Ask yourself: *Am I ready to change?* Because when the answer is yes—when you're tired of the hurt, when you're willing to do whatever it takes to break free—that's when you'll finally move forward.

And when that time comes, you won't just survive. You'll thrive.

EPILOGUE

THE COST OF WISDOM

Many people admire wisdom, but few consider what it takes to earn it. We envy those with unshakable confidence, sharp minds, and the ability to handle anything life throws their way. But would we still desire those qualities if we truly understood the struggles, losses, and sacrifices it took to earn them? So, before we admire someone else's wisdom, we must ask ourselves: are we willing to pay the price it took to earn it?

Because wisdom isn't given—it's earned. And earning it comes at a cost. Sometimes, that cost is our most valuable currency: time. Once spent, you can't get it back. Other times, the cost manifests as pain—mistakes that cut deep, failures that leave you stripped down to nothing, lessons that refuse to be ignored. But more often than not, it's both.

Some lessons in life are straightforward: touch a hot stove, and you get burned. Others are unforgiving, learned only after being knocked down repeatedly until you have no choice but to finally understand. The cost of wisdom is steep. But ignorance costs even more.

Breaking Generational Trauma

Generational trauma doesn't define who we are—it shapes where we start. I was born into a family where pain was passed down like an inheritance. Abuse, addiction, abandonment—these weren't just isolated incidents. They were patterns, cycles that had existed long before I took my first breath.

I wasn't just fighting for my own survival—I was breaking a cycle. The trauma woven through my family's history wasn't mine to carry forward. I made the choice to end it. But that choice came at a cost. Breaking generational trauma isn't just about deciding to do better. It's about years of unlearning what was ingrained in you from birth. It means dismantling patterns, resisting the pull of the familiar, and often standing alone as you carve a new path.

What wisdom did I gain? Healing doesn't just happen. It takes intention, effort, and sacrifice. But breaking the cycle means my daughter will never have to carry the weight of my past. That alone made the cost worth it.

Scars Are Proof You Survived

Some scars fade over time, becoming faint reminders of what once was. Others remain visible, etched into the skin as permanent proof of survival. But the deepest scars? They don't mark the body. They mark the soul.

When I was three years old, I was burned so badly that the doctor feared I'd be permanently disfigured. I didn't understand the gravity of it then, but I remember pressing my face into a red pleather booth, screaming, "It burns!" That scar stayed with me for years, but over time, it faded. What didn't fade was the lesson it carried.

Pain leaves marks, but scars don't mean you're broken. They mean you survived. Every hardship I've endured—every mistake, every loss, every rock-bottom moment—has left a scar. Some are still raw. Others, like that childhood burn, have softened with time. But I no longer see my scars as something to be ashamed of. I see them as proof that I made it.

Emotional Intelligence: A Gift Born from Pain

Some people have the ability to read a room, to sense when something is wrong before a word is even spoken. It's a skill that can make you a great leader, friend, and protector. For me, emotional intelligence wasn't something I was born with naturally—it was something I had to develop to survive.

As a child, I learned to read my father's moods like a second language. I could feel the tension in a room before it exploded. I could sense when danger was near, not because I was taught but because I had no choice. That awareness kept me safe, but it also meant I spent my childhood on edge, constantly monitoring the emotions of others and adjusting my own actions in an attempt to avoid conflict.

That skill followed me into adulthood. At first, it helped me navigate addiction—knowing how to manipulate, how to read what someone wanted to hear before they even said it.

Later on, it became an asset in leadership. I could walk into a room full of people and sense exactly what was happening. I could feel the energy shift before a decision was even made. That ability—to know what needed to be said and how to articulate it—became one of my greatest strengths.

But what was the cost? I had to endure years of abuse to develop it. Emotional intelligence may be a gift, but like many gifts forged in pain, it came at a price no child should ever have to pay. And yet, I found a way to turn it into something powerful.

That instinct, once necessary for survival, evolved into something greater. It enabled me to lead in crises. It allowed me to save lives. It helped me rebuild my own. Not every skill we gain in life comes from a place of privilege or formal education. Sometimes, the sharpest tools are forged through struggle.

For me, emotional intelligence wasn't taught—it was earned. And now, instead of using it to avoid pain, I use it to lead with empathy, to connect, to understand, and to ensure that no one around me feels the way I once did—trapped, unheard, or alone.

The Ripple Effect: Understanding the Costs of Our Actions

Every action we take has consequences—some immediate, while others unfold over many years before we fully understand their impact. The ripple effect is real. Whether motivated by kindness or ego, our choices continue to resonate through time, shaping lives long after we've made them.

I often think of the man who stopped to help me when I needed it most. He had no obligation, no reason to do so—but he did. That single act of kindness created ripples that I've carried with me ever since. It cost him something—his time, his energy—but what he gained was far greater.

Then, there are the ripples of ego. When I skipped English IV and dismissed its importance, I had no idea how that choice would come back to haunt me. Not when I struggled to write in college, nor when I sat down to write this book. Our decisions, whether big or small, all come at a cost. Sometimes, we don't realize the full price we've paid until years later.

Wisdom Through Parenthood

Much of my wisdom comes from pain, failure, and redemption. However, some of the greatest lessons I've learned have come from being a father. Becoming a parent has taught me more than I ever could have imagined. Being responsible for another person's life has shown me the meanings of grace, sacrifice, and selflessness in ways I never knew existed.

But the hardest part? Understanding that wisdom isn't something Ashlynn can inherit—it's something she will have to earn. I know this because I've experienced it. No one could have warned me enough to keep me from making my own mistakes. No lecture, no punishment, no well-intentioned advice could have stopped me from learning things the hard way.

Wisdom isn't handed down. It's paid for. I would give anything to spare my daughter from heartbreak, disappointment, or the pain of learning things the hard way. But I can't protect her from life's lessons any more than I could have protected myself from my own. And as much as every part of me wants to step in, I know that some lessons are hers to learn, not mine to teach. Letting go is hard. Watching her struggle is harder. But at some point, you realize your job as a parent isn't to protect your child from life—it's to prepare them for it.

I didn't just want to break generational trauma—I wanted to give my daughter the tools I never had. I wanted to show her what strength looks like. Not in words, but in action. I wanted her to see that no matter how hard life gets, no matter how many times you fail—you get back up, you keep going, and you stay the course. One day, she will have to pay her own price for wisdom. And when she does, I hope she knows she is strong enough to handle it. Because she comes from a long line of survivors.

The Ability to Handle Extreme Stress

People often say, "I wish I could stay calm under pressure like you." But do they? Do they really want the burden that comes with it? The ability to compartmentalize and think with absolute clarity while others freeze in fear—that kind of composure doesn't come naturally. It's not a gift; it's forged.

I didn't wake up one day knowing how to stay calm in chaos. That skill was instilled in me over years of witnessing suffering, bearing the emotional weight of emergency services, and witnessing death and destruction up close more times than I can count. It began long before I donned a uniform. I learned early how to detach from emotion to survive. As a kid, I learned to read danger before it arrived, to anticipate conflict before it happened. Later, that skill carried into my time in addiction, where compartmentalization became a means of self-preservation—a way to justify, to ignore, to keep functioning in a life that should have killed me.

Then, when I stepped into leadership, that ability became something else. It allowed me to walk into a disaster scene, assess a situation, and make life-or-death decisions without hesitation. It allowed me to lead through the worst situations—traumatic calls, large-scale incidents, and of course the 2016 wildfires —while others stood frozen in fear. But what was the cost? Years of trauma, sleepless nights, and memories that never faded.

Staying calm under pressure is a skill. Some might even call it a gift. But if you want the ability, you have to be willing to pay the price. And that price isn't cheap.

There's a reason so many people crack under pressure—because you never truly know how much stress someone can endure until they are under ***real*** pressure. Some of the toughest people I knew—or thought I knew—buckled when the stakes were life or death. It's easy to appear strong when the hardest decision you make is an inconvenience, but when everything is on the line, when the walls close in, and there's nowhere to run, that's when you see who a person truly is. The burden of carrying that weight, of never allowing yourself to break, of always being the steady one in the storm—it takes its toll. Not everyone can handle it. And once you've been tested, you never look at "toughness" the same way again.

But would I change it? Not for a second. Because when the moment comes, when others are looking to you to lead, you don't have the luxury of falling apart. You stand. You lead. You stay the course.

The Most Important Lesson of All: No One Is Coming

For years, I carried my past like a weight on my back. The harder I tried to outrun it, the heavier it became. I spent my early adulthood searching for an easy way out—waiting for someone or something to come along and fix what was broken inside me. I believed that if I could just find the right person or the right distraction, then maybe, just maybe, I could escape the mess I had created. But life doesn't work that way.

At my lowest, when I was strung out, homeless, and completely lost, I learned the hardest, most valuable lesson of my life: *No one was coming to save me.*

There was no lifeline, no shortcut, no second chance waiting for me if I just sat still long enough. The people I had once relied on were either gone, fed up, or broken themselves. I had burned too many bridges and exhausted too many opportunities. And as much as I wanted to believe that someone would pull me out of the darkness, no one could—because no one could do the work for me.

That realization struck harder than any withdrawal or rock-bottom experience. It was both terrifying and liberating. If no one was coming, it meant everything was up to me—and for the first time, I viewed that as an opportunity rather than a punishment. If I wanted my life back, I had to fight for it.

That fight didn't happen all at once; it was slow and grueling. It was ugly, unglamorous, and often thankless. It was about making one choice at a time, putting one foot in front of the other, again and again, until those small steps formed a foundation. I didn't just rebuild my life—I created a new one.

Many people never learn this lesson. Some spend their whole lives waiting for a rescue that will never come. They convince themselves that the rescue lies in a better relationship or a stroke of luck. But the truth is: *the only way out is through.* The only way forward is *to stay the course.*

Staying the Course

The easiest thing to do when life gets hard is to quit. The hardest thing to do is to keep going when there's no guarantee it will get better.

Every lesson I've learned and every hardship I've faced has tested my ability to stay the course. There were moments when I wanted to give up, times when I thought the pain wasn't worth the fight. Days when the idea of just "existing" seemed easier than the effort of trying to rebuild. But wisdom doesn't come to those who stop halfway. It comes to those who push forward, even when the road ahead is uncertain.

When I got clean, it wasn't just about quitting drugs—it was about learning how to live without them. How to cope, how to face the thoughts I had spent years suppressing, how to retrain my entire brain to function without something artificial keeping me afloat. When I fought through PTSD, it wasn't about trying to erase the trauma—it was about learning how to carry it without letting it crush me. And when I rebuilt my life after every setback, I stayed the course. That's what made the difference.

Anyone can start over. The real test is whether you can keep going when everything in you wants to stop. Some days, staying the course meant nothing more than getting out of bed and doing the next right thing—not because I felt like it, but because I refused to let failure be the end of my story. That is what separates those who rebuild from those who remain lost.

Staying the course isn't about perfection. It's about persistence. And if there's one thing I know for sure, it's this: No matter how broken you feel, no matter how impossible the road ahead seems—you can keep going. You must keep going. Because in the end, the only way to rebuild a life is to refuse to give up on it.

The Final Lesson: The Cost of Wisdom

I once stood in a bathroom, staring at a man I didn't recognize – a man strung out and broken. While my mind screamed the question, I didn't have an answer for: *How the fuck did I get here?* I felt trapped, as if every mistake I'd ever made had brought me to that single moment. Shame, regret, and fear—they were all I knew. That day, I could hardly look at myself in the mirror.

But today? Today, I stand in front of a mirror and see a man who fought to earn every piece of wisdom he carries. He is not a victim of his past. He is the architect of his future. And if I had to pay the price all over again—I would. Because the cost of wisdom is high, but the price of ignorance is higher.

If there's one thing I want you to take away from my life, it's this: No one is too far gone. No one is beyond redemption. No one is past healing. No matter how heavy the burden is, no matter how far you've fallen – you can rebuild. There is hope waiting on the other side. If you're struggling right now, if life feels heavier than you can bear, remember this—every step forward matters. Stay the course. The road is hard. The cost is high. But on the other side of struggle is wisdom, strength, and a life you fought to rebuild.

ACKNOWLEDGMENTS

This book would not exist without the love, grace, and support of those who walked beside me through every step of this journey—whether they knew it or not.

To my daughter, Ashlynn: You are my greatest teacher, my greatest joy, and the reason I *stayed the course* when everything in me wanted to give up. Watching you grow into the strong, compassionate woman you are today has been the most rewarding part of my life. This book, and everything it stands for, is for you.

To my family—especially my mom—for believing in me even when I couldn't believe in myself. Your prayers, sacrifices, and quiet strength have laid the foundation for my recovery.

To Larry, Cindy, and Greg: Thank you for trusting me with responsibility when others might have turned away. Your belief in me made a significant difference at a critical point in my life, and I will never forget your support.

To David—my friend and brother in every way except blood—thank you for supporting me, especially when the burden of leadership felt too heavy to bear alone. Your memory continues to inspire me to honor the promises we made.

To Meghan Davis of *Tell More Stories*—thank you for helping me bring this vision to life. Your patience, talent, and ability to transform my raw memories into something meaningful have been a gift for which I will always be grateful. You didn't just assist me in writing a book—you helped me convey my truth.

ENDNOTES

[i] Baker, V. (2015). Battle of the bans: US author Judy Blume interviewed about trigger warnings, book bannings and children's literature today. Index on Censorship. https://doi.org/10.1177/0306422015605719a

[ii] Prevent Child Abuse America. (n.d.). Donna J. Stone Award. Prevent Child Abuse 50. Retrieved January 23, 2025, from https://preventchildabuse50.org/#:~:text=Donna%20J.,on%20child%20abuse%20and%20neglect

[iii] Florida Department of Law Enforcement. (n.d.). Drug Abuse Resistance Education (D.A.R.E.) program overview. Retrieved August 10, 2024, from https://www.fdle.state.fl.us/CJSTC/DARE.aspx

[iv] "Opioid Facts and Statistics," U.S. Department of Health and Human Services, accessed September 2, 2024, https://www.hhs.gov/opioids/statistics/index.html

[v] "Causes of Addiction," Recovered, accessed September 2, 2024, https://recovered.org/addiction

[vi] U.S. Department of Justice. (n.d.). *Background information on OxyContin.* National Drug Intelligence Center. Retrieved from https://www.justice.gov/archive/ndic/pubs/651/backgrnd.htm

[vii] Time Magazine. (2019, November 11). The Science of Addiction.

[viii] Volkow, N.D., Koob, G.F., & McLellan, A.T. (2016). *Neurobiological advances from the brain disease model of addiction.* The New England Journal of Medicine, 374(4), 363-371. https://doi.org/10.1056/NEJMra1511480

[ix] U.S. Food and Drug Administration. (n.d.). *Developing a risk prediction engine for relapse in opioid use disorder*. U.S. Food and Drug Administration. https://www.fda.gov/science-research/advancing-regulatory-science/developing-risk-prediction-engine-relapse-opioid-use-disorder#:~:text=In%20the%20United%20States%2C%202.7,rates%20of%2065%2D70%25

[x] Substance Abuse and Mental Health Services Administration. *Medication-Assisted Treatment: History and Evolution.* SAMHSA, 2015.

[xi] National Institute on Drug Abuse. "How Methadone Works to Treat Opioid Use Disorder." NIDA, 2020.

[xii] Mayo Clinic. "Buprenorphine vs. Naltrexone for Opioid Dependence: Key Differences." Mayo Foundation for Medical Education and Research, 2019.

[xiii] Data collected from the After-action Review of the 2016 Gatlinburg Wildfire

[xiv] 445 fire trucks and 3,500 responders: According to an article by NBC News, published on December 4, 2016, titled "Gatlinburg Fire Officials: Death Toll Climbs to 14, 1,000 Structures Damaged," the wildfire had prompted the deployment of 445 fire trucks and more than 3,500 responders to combat the blaze.

[xv] 3,000 people sheltered: According to an article by CNN, published on December 2, 2016, titled "Gatlinburg fire: 3 dead; 14,000 evacuated; 700 buildings damaged," around 3,000 people had been sheltered in the aftermath of the wildfire.

[xvi] 20,000 volunteers and 16,000 meals served: According to an article by WATE 6 On Your Side, a local news station, published on January 6, 2017, titled "Dollywood Foundation to give $10K to all families who lost their home in Sevier County wildfires," approximately 20,000 volunteers had stepped up to assist with the disaster relief efforts, and over 16,000 meals had been served.

[xvii] According to an article by The Tennessean, published on December 15, 2016, titled "Gatlinburg fire recovery gets $12M in donations from Dolly Parton telethon," the "Smoky Mountains Rise: A Benefit for the My People Fund" telethon hosted by Dolly Parton had raised over $9 million, with additional donations bringing the total to $12 million.

[xviii] USA Today: "Lawsuit against boys blamed for 2016 Gatlinburg fire dismissed" (December 14, 2019)

[xix] WBIR News: "Mixed emotions after lawsuit against teens accused cf starting Gatlinburg wildfires is dismissed"

[xx] Substance Abuse and Mental Health Services Administration (SAMHSA). (2020). Key Substance Use and Mental Health Indicators in the United States: Results from the 2019 National Survey on Drug Use and Health (HHS Publication No. PEP20-07-01-001, NSDUH Series H-55). Rockville, MD: Center for Behavioral Health Statistics and Quality, Substance Abuse and Mental Health Services Administration.

[xxi] American Foundation for Suicide Prevention. (2021). Suicide statistics.

[xxii] Oliffe, J. L., & Phillips, M. J. (2008). Men, depression, and masculinities: A review and recommendations. Journal of Men's Health, 5(3), 194–202. Wong, Y. J., Ho, M. R. A., Wang, S. Y., & Miller, I. S. K. (2017). Meta-analyses of the

relationship between conformity to masculine norms and mental health-related outcomes. Journal of Counseling Psychology, 64(1), 80–93.

[xxiii] Draganski, B., & May, A. (2008). Training-induced structural changes in the adult human brain. Behavioural Brain Research, 192(1), 137-142.

[xxiv] Diamond, M. C. (2001). Response of the brain to enrichment. Annual Review of Psychology, 52, 1-26.

Made in the USA
Columbia, SC
26 April 2025